IET COMPUTING SERIES 56

Earth Observation Data Analytics Using Machine and Deep Learning

Other volumes in this series:

Earth Observation Data Analytics Using Machine and Deep Learning

Modern tools, applications and challenges

Edited by
Sanjay Garg, Swati Jain, Nitant Dube and
Nebu Varghese

The Institution of Engineering and Technology

Published by The Institution of Engineering and Technology, London, United Kingdom

The Institution of Engineering and Technology is registered as a Charity in England & Wales (no. 211014) and Scotland (no. SC038698).

The Institution of Engineering and Technology
Futures Place
Kings Way, Stevenage
Hertfordshire SG1 2UA, United Kingdom

www.theiet.org

British Library Cataloguing in Publication Data
A catalogue record for this product is available from the British Library

ISBN 978-1-83953-617-5 (hardback)
ISBN 978-1-83953-618-2 (PDF)

Typeset in India by MPS Limited

Cover Image: Digital Earth, conceptual illustration / EDUARD MUZHEVSKYI / SCIENCE PHOTO LIBRARY via Getty Images

Contents

About the editors

Sanjay Garg is a professor in the Department of Computer Science and Engineering at the Jaypee University of Engineering and Technology, India. His research interests include data science, algorithms and pattern recognition. He has over 30 years of academic and research experience. He has completed six funded research projects sponsored by ISRO under the RESPOND scheme and GUJCOST as principal investigator in the field of earth observation data analytics and image processing. He has supervised 10 doctoral dissertations in the same field. He is a fellow of the Institution of Engineers (India), a senior member of IEEE and a Senior member of ACM.

Swati Jain is an associate professor in the Department of Computer Science and Engineering at the Institute of Technology, Nirma University, India. She works in the areas of machine learning, data analytics, and deep learning. She is currently working on four funded research projects, two of them under BRNS Department of Atomic Energy and two by ISRO (Indian Space Research Organization). She has completed one funded research project under ISRO-RESPOND as a CO-PI. She is actively working on establishing a Center of Excellence in-Data Science in association with Binghamton University, USA. She received her PhD degree from Nirma University in the areas of machine learning and image processing.

Nitant Dube is the group director of MOSDAC Research at the joint Space Applications Centre (SAC) and ISRO, India. His research fields include satellite image processing, big data analytics, AI/ML and its applications for Earth observation data, geo-intelligence and web-based processing. He is involved in the design and development of software for meteorological and oceanographic applications. He has been responsible for the design and development of data products and information processing systems for Indian remote sensing satellites and has contributed towards the development and operationalization of data processing systems at Indian and International ground stations. He is an ISRO nominated member for the CEOS Working Group on Information System and Services (WGISS). He holds a PhD degree in Computer Science from Nirma University, Ahmedabad (GJ), India.

Nebu Varghese is an assistant manager (GIS) in the Land and Municipal Service at Dholera Industrial City Development Limited (DICDL), India. He works in the areas of GIS systems analysis and prepare design for new GIS methodologies, land use mapping, land cover mapping, urban land use analysis, spatial data management, satellite image processing and analysis, machine learning and deep learning. Currently, he is working on regional and city-level planning projects, where he employs the most cutting-edge technologies for building information model (BIM) to GIS Integration with micro-level asset information of all Infrastructure in city development. He has been involved in various government DST, ISRO, and IIMA funded projects and was also involved in the innovation hub Malawi project with the Deutsche Gesellschaft für Internationale Zusammenarbeit (GIZ), India. He is a member of ISPRS. He holds a master's degree in remote sensing & GIS from Sam Higginbottom University of Agriculture, Technology & Sciences (SHUATS), Prayagraj (UP), India.

Foreword

This compilation of various research outcomes in the form of a book is an excellent collection of well-written extensive papers on various important and current topics of application of modern computational technologies in the field of image processing of remote sensing data. These are outcomes of their own studies and research on various topics. The topics covered range from Geospatial Big Data Analysis Using Neural Networks Deep Learning methods for Crop classification, Transfer Learning Approach for Hurricane Damage Assessment, Wildfires, Volcanoes and Climate Change, Monitoring Using Deep Neural Networks, Exploiting Artificial Immune Networks for Enhancing Remote Sensing Image Classification, Mining of Earth Observation Data, and Software Framework for Spatiotemporal Data Analysis. The authors have taken a lot of care in reviewing an extensive list of literature on the topics covered. All these topics are of current intensive research.

Usually, there is a strong urge to bring about drastic changes that are urgently required to move the world toward a sustainable and resilient course. To attain these goals, fast computing methods and sophisticated algorithms for pattern recognition are need of the hour. Various machine learning and deep learning algorithms have these potentials which are needed to be exploited. Understanding the scope and implications of this global challenge will require global approaches. Only global information that is easily accessible from affordable sources, such as satellite images and other widely available sources, can help us achieve sustainable goals through their use and ensure their universality. Original and precise methods for addressing the indicators linked to sustainable development goals are the need of the hour.

A plethora of Earth-observing satellites gathers enormous amounts of data from various sources. But today, the challenges in effectively utilizing such data lie in data access technologies in various data formats, data visualization, and various data processing methodologies, computational speeds. If the solutions to these issues can be found, the scientific communities would have faster access to much better data to support decisions making on weather, climate forecasting and their impacts, including high-impact weather events, droughts, flooding, wildfires, ocean/coastal ecosystems, air quality, and many more.

This book is a comprehensive resource that brings together the most recent research and techniques to cover the theory and practice of Earth Observation data analytics. The editors and authors have long experience in their fields and have put their knowledge and expertise in various fields together in seeking solutions to various problems in earth observation systems.

I strongly recommend this book to all researchers and students interested in these fields. The chapters in the book will certainly go a long way in enhancing their knowledge and expertise in the files of their study. Certainly, not least, this book will be a valuable and timely addition to both their personal collections as well as their intuitional libraries.

Dr M.B. Potdar
Former Scientist (Indian Space Research Organization)
ISRO/Space Applications Centre, Govt. of India

Former Project Director (Bhaskaracharya National Institute for
Space Applications and Geo-informatics) BISAG, Govt. of Gujarat.

Chapter 1

Introduction

Preeti Kathiria[1], Swati Jain[1], Kimee Joshi[1] and Nebu Varghese[2]

1.1 Earth observation data

Sustainable development and climate change are problems that require immediate solutions, and both are critical to humanity's present and future well-being [1,2]. Also, human activities are increasing strain on natural resources, which has a global impact on the environment. Continuous and ongoing monitoring is required to analyze, comprehend, and minimize these environmental changes [3]. The United Nations (UN) seeks a sustainable development model for this generation and future generations, as well as shared prosperity for people and the planet, through the promotion of its Sustainable Development Agenda and the United Nations Framework Convention on Climate Change (UNFCCC) [4,5]. The UN has defined a set of 17 sustainable development goals (SDGs) as a plan of action to reach peace and prosperity for all people on our planet by 2030. Several benchmarks and indicators for each of the 17 goals are used to measure, track, and report the development of every nation. The global framework established by the UN is designed around 169 targets, and 232 indicators, 71 (42%) of these targets and 30 (13%) of the indicators can be measured directly or indirectly by Earth observation (EO) [6–8].

EO plays an essential role in advancing many of the SDGs. Addressing scientific issues like global warming and climate change, ecological change, and reduction effects of habitat and biodiversity deterioration and producing statistics and indicators that allow the quantification of SD. The UN report has shown the viability of using EO data to produce official statistics, including SDGs statistics like agricultural, urban, and land planning, or food security indicators [9]. Data on the state of the atmosphere [10], oceans [11], crops [12], forests [13], climate [14], natural disasters [15], natural resources [16], urbanization [17], biodiversity [18], and human conditions [19] can be provided by EO. The SDGs that benefit from all EO indicators are zero hunger (SDG 2), clean water and sanitation (SDG 6),

[1]Institute of Technology, Nirma University, India
[2]Dholera Industrial City Development Limited, India

climate action (SDG 13), life below water (SDG 14), and partnership for the goals (SDG 17). EOs from satellites and airborne and in situ sensors provide accurate and reliable information on the state of the atmosphere, oceans, coasts, rivers, soils, crops, forests, ecosystems, natural resources, ice, snow, and building infrastructure it changes over time. These observations are directly or indirectly required for all governmental functions, all economic sectors, and nearly all daily societal activities [7]. EO satellites make real-time observations of the land, ocean, atmosphere, cryosphere, and carbon cycle from space, which continuously relay this data to the ground.

1.1.1   Organization

This paper presents EO data along with various applications. The rest of the paper is organized as follows: Section 1.2 discusses the categories of the EO data, Section 1.3 describes the need of data analytics, Section 1.4 describes the data analytics methodology, Section 1.5 shows a data visualization techniques, Section 1.6 presents the application areas, and concluding remarks are in Section 1.7.

1.2   Categories of EO data

This section proposes various types of EO imagery. The process of collecting observations of the Earth's surface and atmosphere using remote sensing tools is known as Earth Observation. The captured data is typically in the form of digital images [20]. The primary differences are based on the sensing device used to capture the image (passive or active) and the wavelength of the electromagnetic spectrum used for the observation. They are broadly classified into two categories as shown in Figure 1.1: passive imaging system and active imaging system.

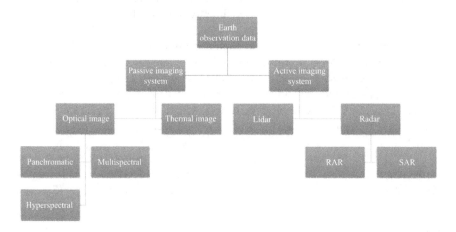

Figure 1.1   Overview of an EO data

1.2.1 Passive imaging system

Starting with the passive imaging system, passive EO sensors either measure the thermal IR or microwave radiation emitted from the Earth or measure solar energy reflected by the Earth's surface in the visible to the middle infrared region of the electromagnetic spectrum [21]. Passive sensors that are frequently used in EO will be introduced in the subsections in the following categories.

1.2.1.1 Optical image

Optical sensors detect and record radiation in the visible, near-infrared, and short-wave wavelengths for either:

1. One channel (panchromatic)
2. Several channels (multispectral)
3. Numerous channel (hyperspectral)

Panchromatic
Panchromatic scanners typically capture electromagnetic radiation (EMR) in a single band that includes all wavelengths from the visible to infrared spectrum. A grey-scale image, which makes pixels with lower image values appear dark and those with higher values appear bright, is the most common way to display panchromatic data. Concerning optical imagery, panchromatic channels typically record low values for water and dense vegetation and high values for urban and bare areas [21]. Panchromatic sensors produce images with higher spatial resolution than multispectral scanners [20,21].

Multispectral
The creation of "natural colour" images using measurements from three visible spectrum bands is a typical example of a multispectral image (narrow bands centered around the blue, green, and red wavelengths) [20]. As the name suggests, multispectral scanners (MSS) are a specific form of remote sensing equipment that detects and digitally records radiation in various, defined wavelength areas of the visible and infrared parts of the electromagnetic spectrum [21]. Multi-spectral instruments typically have to collect energy on larger spatial extents to "fill" the imaging detector, resulting in a lower resolution than for panchromatic images because the range of wavelengths contributing to the radiation energy detected by the sensor is reduced.

Hyperspectral
Hyperspectral scanners collect image data for hundreds of spectral channels. Instead of assigning primary colours (red, green, blue) to each pixel, hyperspectral imaging (HSI) analyzes a broad spectrum of light. To provide more information on what is imaged, the light striking each pixel is broken down into many different spectral bands [22].

In Figure 1.2, the images a and b handled by [23,24] are shown here as an example of a panchromatic image obtained from SPOT satellite with 10 m resolution and a multispectral image obtained from plants cope with 3.7 m resolution,

Figure 1.2   (a) Panchromatic image. (b) Multispectral image. (c) Hyperspectral image.

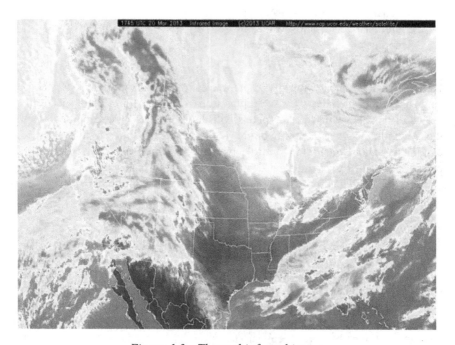

Figure 1.3   Thermal infrared image

respectively. In the same figure, image c shows the picture of a portion of Horseshoe Bay Village in Xiong' with 0.5 m spatial resolution reported by the Institute of Remote Sensing and Digital Earth of the Chinese Academy of Sciences and the Shanghai Institute of Technical Physics of the Chinese Academy of Sciences [25].

1.2.1.2   Thermal image

Thermal infrared radiometers detect energy emitted from the Earth's surface with wavelengths ranging from 3 to 15 m [21]. Since the infrared spectrum is not visible to the human eye, thermal infrared imaging is also known as "non-visible" imaging [26]. With high thermal and spatial resolutions, thermal imaging maps any object's surface temperature [27]. The image shown in Figure 1.3 is downloaded from [28,29].

1.2.2 Active imaging system

Active remote sensing devices work on the principle of transmitting energy, either as pulses or as a continuous signal, towards a specific target, then measuring the energy returned from the target. Radar and lidar are two different subcategories of active imaging sensors [21].

1.2.2.1 Lidar

Light detection and ranging (lidar) is a technique that uses a transmitted laser pulse to detect the presence of a target and measures the distance to the target based on the time and intensity of any reflected return pulse. Lidar is an active remote sensing technology that operates in the ultraviolet to near-infrared wavelength range. Lidar systems are available on a variety of sensing platforms, including satellite, airborne, and ground-based systems [21].

1.2.2.2 Radar

RADAR is an acronym for RAdio Detection And Ranging, that is, using actively transmitted radio waves to detect objects and determine their position or 'range'. For EO, the X, C, and L bands are the most commonly used [21]. Compared to infrared and optical sensing devices, the primary goal of radar is to detect distant targets under adverse weather conditions and determine their distance, range, and precision. The radar has a transmitter that serves as an illumination source for target placement. It generally operates in the microwave region of the electromagnetic spectrum, which is measured in Hertz [30]. The electrical wave vibrations in the transmitted radar signals can be constrained to a single plane, that is, perpendicular to the wave propagation direction (rather than vibrating in all directions perpendicular to that of propagation). The term "polarization" refers to this filtering procedure. In imaging radar, there are two orthogonal polarization modes known as horizontal (H) and vertical (V), which are normally transmitted individually [21]. Two types of radar-based systems are commonly used for microwave imaging on aircraft and satellite platforms:

1. Side-Looking Airborne Radar (SLAR) or Real Aperture Radar (RAR)
2. Synthetic Aperture Radar (SAR)

SLAR stands for synthetic aperture radar (Side Looking Airborne Radar). The illumination of both Real Aperture Radar and Synthetic Aperture Radar is typically perpendicular to the flight path, making them side-looking systems. The along-track resolution, also known as the azimuth direction, distinguishes one system from another. Actual Aperture Radars' Azimuth resolution is based on the antenna beam width, which is proportional to the distance between the radar and the target (slant range). Using a series of signals that have been stored in the system memory, synthetic aperture radar uses signal processing to create an aperture that is hundreds of times more significant than the actual antenna [31].

SLAR was the first active sensor to produce terrain imagery from back-scattered microwave radiation. An antenna is mounted beneath the platform in SLAR to produce a fan beam (wide vertically and narrow horizontally) pointing to the platform's side [32].

Figure 1.4   SAR image

Synthetic aperture radars were developed to overcome the limitations of real aperture radars. For good Azimuth resolution that is independent of the slant range to the target, these systems combine small antennae with relatively long wavelengths [31]. The image shown in Figure 1.4 is downloaded from [33] Copernicus Open Access Hub.

1.3   Need of data analytics in EO data

EO data analytics has been regarded as a major challenge since the dawn of time. The amount of data available today is exponentially increasing, but it is only partially structured and harmonized. Understanding the scientific, socioeconomic, and environmental benefits of earth observation data analytics has become critical for businesses and users alike [34,35]. The primary goal of data analytics is to uncover hidden patterns, unknown correlations, and other useful information from large amounts of heterogeneous data to aid in Earth science research [36].

1.4   Data analytics methodology

Data analysis is the process of transforming raw data into something useful and comprehensive for a specific purpose. Following preprocessing, the primary goal of data analytics is to elucidate obscured patterns, unidentified correlations, and other pertinent information from a sizable volume of heterogeneous data to support Earth science research [36,37].

Table 1.1   Data analytics method

Analytics type	Methods
Machine learning	Classification [18], Clustering [12], Regression [38], Dimension reduction [39]
Deep learning	Classification [40,41], Object detection [42], Image segmentation [42]

Table 1.2   Examples of EO satellite applications

Field	Main findings	Method	References
Agriculture	The method was successful in mapping the dynamics of crop types in complex landscapes with small field sizes.	Random Forest	[43]
	Combining MODIS time series data with machine learning algorithms allows for the detection of successive crops.	Random Forest	[44]
Land cover	Classify RISAT-1 dataset over the Mumbai region for land Cover classification.	CNN	[45]
Maritime	The approach was used to detect and map terrestrial oil contamination.	Random Forest	[46]
	Developed an Oil Spill Detection framework based on a deep learning algorithm.	CNN	[47]
	Develop an algorithm for automatic ship detection from SAR and Optical ship dataset.	Improved-YOLOv3	[48]

To identify the ambiguous and complex relationships between variables and to better comprehend the geographic distribution and frequency distribution of big Earth data, traditional statistical methods, which are frequently predicated on specific assumptions, are frequently used [36]. In terms of non-linear relationship understanding, machine learning methods generally outperform traditional statistical methods [36]. The involved methods can be categorized as machine learning and deep learning (Table 1.1).

1.4.1   Machine learning

Machine learning is a branch of computational algorithms that is constantly developing and aims to mimic human intelligence by learning from the environment [49]. Machine learning is broadly classified into three subdomains: supervised learning, unsupervised learning, and reinforcement learning. To summarize, supervised learning necessitates training on labeled data with inputs and desired outputs. Unsupervised learning, as opposed to supervised learning, does not require labeled training data, and the environment only provides inputs with no desired targets. Reinforcement learning allows for learning based on feedback from interactions with the outside world [50].

Machine learning classifiers including Random Forest, Support Vector Machines, and Maximum likelihood classifiers can produce the probability of an observation belonging to a specific class of Earth process, such as land use and land cover classification [51], and crop classification [52,53].

1.4.2   Deep learning

Deep learning techniques, which emerged from machine learning, have unique capabilities for extracting and presenting features from Earth data at various, detailed levels. In the classification and segmentation of Earth data tasks, these features and characteristics are crucial. Deep learning has demonstrated excellent performance in a variety of fields, including natural language processing, computer vision, recommendation systems, and others, thanks to its more potent expression and parameter optimization capabilities [54–56]. For example, the deep convolutional neural networks (CNNs), ResUnet, and DeepUNet can perform satisfying results in Earthquake-triggered Landslide detection [41,57] classified the different types of the crop from multi-temporal data. Marmanis *et al.* [40] classified different types of the scene using CNN classifier. Beyond image classification, objects can be detected and segmented using deep learning techniques [42,58].

1.5   Data visualization techniques

A graphic representation of information and data is referred to as data visualization [59]. The following is a list of different types of map layers or map types that are mostly used to visualize data [60,61]. Figure 1.5 shows different types of map [60,61] to visualize the data.

1.5.1   Cartogram map

This choropleth map variant combines a map and a chart. It entails taking a land area map of a geographical region and segmenting it so that the sizes and/or distances are

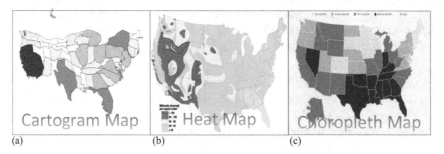

(a)   (b)   (c)

Figure 1.5   (a) Cartogram map [62]. (b) Heat map [63]. (c) Choropleth map [64].

proportional to the values of the variable being measured. Then, to correspond to its corresponding value, each segment is assigned a different colour or shade. As a result, the data is more directly related to the land area to which it refers [60,61].

1.5.2 Heat map

Heat map uses colors or shades to represent various values or value ranges. A heat map differs from a choropleth map in that its colors do not correspond to geographic boundaries [60]. Instead of presenting these values and ranges as discrete cells constrained by arbitrary geographic or political boundaries, it presents them as a continuous spectrum.

A heat map can help you more clearly to see patterns of high ("hot spots") and low concentrations of a variable in this way. However, since algorithms are frequently used to transform discrete data points into a continuous spectrum, this can compromise accuracy.

1.5.3 Choropleth map

One more common kind of map is a choropleth map. It is created by first dividing the area to be mapped into sections, such as by political or geographical boundaries, and then filling each section with a different color or shade. Referring to [61], every shade or color represents a different value or range of values that a variable can have [60].

1.6 Types of inferences from data analytics (application areas)

Figure 1.6 shows the different application areas of Earth Observation data such as Agriculture, Forestry, Flooding, Land Cover Classification, Maritime, Wetland, Defence and Security.

1.6.1 Agriculture

For planning and decision-making objectives such as distribution and storage of food grains, governmental policies, pricing, procurement, and food security, among others, it is necessary to have access to crop statistics. The Ministry of Agriculture and Farmers' Welfare successfully employs modern satellite remote sensing technology in such decision-making [65].

Satellite imagery has the potential to increase revenue generation for agricultural applications by providing information related to crop type, crop insurance damage assessment, production management techniques, fertilizer application requirements, yield estimations, re-growth monitoring, illicit crop monitoring, pest and invasive species monitoring, and irrigation requirements and application, monitoring agri-environmental measures (such as acreage) to inform subsidy allocations, field boundary management, crop health mapping, field scale mapping, and storm damage assessment [66] (see Figure 1.7).

1.6.2 Forestry

Over 1.6 billion people rely on forests for their livelihoods and their sources of food, medicine, and fuel. Forests cover 31% of the total area of the planet. Provide

Figure 1.6   Applications of EO

more information for Obtaining information on forest acreage, stand density, species composition, age, and condition to be recognized as a For management purposes, a single unit) surveying, assessing, and keeping track of forest health,
Updating forest management plans: tree cutting, delineation, and tracking of par-
cels, estimating biomass, assessing plant health, and plantation surveillance,
Estimating damage from fire, storms, and other extreme weather, Conservation
area planning and protection, "Conducting fuel analysis and locating the locations
where the Fire risk is high, Deforestation mapping, Monitoring of forest conservation and regrowth initiatives [66] (see Figure 1.8).

1.6.3   Land cover classification

Land can be classified, allowing us to learn about its various uses and track its
evolution over time. This is useful for a variety of applications, including monitoring mining and resource extraction, measuring deforestation, land cadastre, and
planning regulations [66].

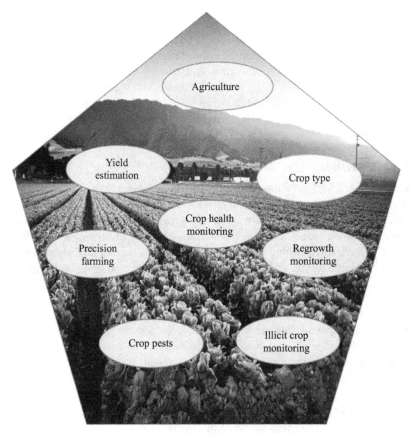

Figure 1.7 Application areas of agriculture

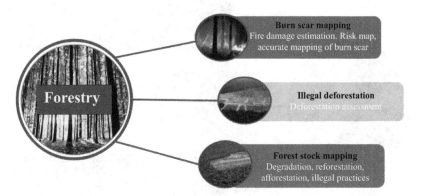

Figure 1.8 Application areas of forestry

1.6.4   Flooding

Monitoring soil moisture and water levels, and how they fluctuate over time provide a solid indicator of how likely it is that flood and drought threats may occur. Combining high fidelity interferometric synthetic aperture radar (SAR) measurements with ground truth measurements, S-band SAR, medium resolution optical images, and digital elevation modeling can be used to achieve this. Combining this with accurate weather forecasts to determine the likelihood and anticipated amount of rain enables the detection of potential flood and drought concerns [66].

1.6.5   Maritime

Oceans comprise 96.5% of the water on Earth and makeup 70% of its surface. Ship tracking data from bathymetry is used to create nautical charts and measure beach erosion, subsidence, and sea levels. Detection of iceberg threats on shipping routes, protecting the environment in the ocean, oil spill monitoring, detection of an illegal oil discharge, detection of unlicensed fishing vessels, port surveillance, detection of incoming hostile objects in maritime piracy, and marine environmental protection [66] (see Figure 1.9).

1.6.5.1   Ship detection

Radar and optical imagery are both useful for tracking maritime activity and detecting ships. These EO systems' adaptability will enable the detection, type-classification, and movement monitoring of ships, boats, and other vessels. In addition to finding ships, it is also possible to determine other details like speed, heading, and, depending on resolution, the general class of ship. Law enforcement, including enforcing regulations governing fishing activities, environmental protection, search and rescue, ship traffic monitoring, as well as customs and excise activities like preventing illegal smuggling activities, all make use of ship detection information [66].

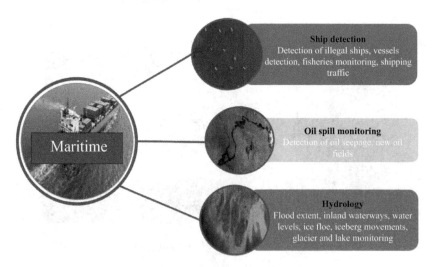

Figure 1.9   Application areas of Maritime

1.6.5.2   Oil spill monitoring

In the oceans across the world, many million tonnes of oil are spilled annually. One of the major ocean pollutants is an oil spill, which poses a serious threat to the marine environment [47]. Most maritime and coastal nations require the ability to detect and monitor oil slicks on the sea's surface to enforce maritime pollution laws, identify offenders, support cleanup and control efforts, detect oil spills from far-off pipelines, and detect oil seepage from ocean floors, which could indicate the presence of new oil fields. SAR photographs can identify oil slicks because of the way they differ from oil-free regions in appearance [66].

1.6.5.3   Hydrology

A hydrological survey is essential for understanding the coastal zones and inland waterways of a territory. Timely and reliable assessments of available water resources via satellites and models provide critical input for developing strategies and water management in the country [67]. Coastal and tidal zones that change frequently due to meteorological influences can be of concern. Using satellite technology, information such as shallow water depths, mudflat topology, and the presence or absence of outflow or sediments can be derived. Satellite imagery provides information on changing bathymetry in coastal zones, which is especially useful around ports and busy shipping areas. The surveys that can be conducted from space provide a consistent and accurate overview [66].

1.6.6   Defence and security

Imaging capability aids nations seeking to better understand and police their borders, coastlines, and assets. Monitoring the flow of people and goods into and out of a country helps policymakers make better decisions.

1.6.7   Wetland

Wetlands are the most productive ecosystems, with a diverse range of flora and fauna. Wetland conservation is therefore critical for preserving biological diversity. Wetlands are currently under stress due to biological diversity loss, deteriorating water quality, sedimentation and shrinkage in the area, infestation by unwanted weeds, and other factors. Remote sensing data is the primary source of information for monitoring and mapping large areas, such as wetland extent, distribution, and wetland types such as freshwater, peat swamps, and non-forested peatlands, among others [68].

1.7   Conclusion

In this paper, we have presented detailed information on EO data and various types of EO data. Then, we have listed various data analytics methods used to represent the data into meaningful manner. Then, we have listed some data visualization techniques. Finally, this paper presented types of inferences from data analytics.

References

[1]   Steffen W, Richardson K, Rockström J, *et al*. Planetary boundaries: guiding human development on a changing planet. *Science*, 2015;347(6223):1259855; 2015.

[2]   Biermann F, Bai X, Bondre N, *et al*. Down to earth: contextualizing the Anthropocene. *Global Environmental Change*. 2016;39:341–350.

[3]   Giuliani G, Camara G, Killough B, *et al*. Earth observation open science: enhancing reproducible science using data cubes. *MDPI*. 2019;4:147.

[4]   Transforming Our World: The 2030 Agenda for Sustainable Development, 2015. Available from: https://undocs. org/en/A/RES/70/71.

[5]   Paris Agreement, 2015. Available from: https://unfccc.int/process-and-meetings/the-paris-agreement/theparis- agreement.

[6]   Persello C, Wegner JD, Hänsch R, *et al*. Deep learning and earth observation to support the sustainable development goals: current approaches, open challenges, and future opportunities. *IEEE Geoscience and Remote Sensing Magazine*. 2022;10(2):172–200.

[7]   Earth Observations in Support of the 2030 Agenda for Sustainable Development. Available from: https://www.earthobservations.org/documents/publications/201703_geo_eo_for_2030_agenda.pdf.

[8]   Estoque RC. A review of the sustainability concept and the state of SDG monitoring using remote sensing. *Remote Sensing*. 2020;12(11):1770.

[9]   Ferreira B, Iten M, and Silva RG. Monitoring sustainable development by means of earth observation data and machine learning: a review. *Environmental Sciences Europe*. 2020;32(1):1–17.

[10]   Boyte SP, Wylie BK, Howard DM, *et al*. Estimating carbon and showing impacts of drought using satellite data in regression-tree models. *International Journal of Remote Sensing*. 2018;39(2):374–398.

[11]   Poursanidis D, Topouzelis K, and Chrysoulakis N. Mapping coastal marine habitats and delineating the deep limits of the Neptune's seagrass meadows using very high resolution Earth observation data. *International Journal of Remote Sensing*. 2018;39(23):8670–8687.

[12]   Reza MN, Na IS, Baek SW, *et al*. Rice yield estimation based on K-means clustering with graph-cut segmentation using low-altitude UAV images. *Biosystems Engineering*. 2019;177:109–121.

[13]   Puletti N, Chianucci F, Castaldi C, *et al*. Use of Sentinel-2 for forest classification in Mediterranean environments. *Annals of Silvicultural Research*. 2018;42(1):32–38.

[14]   Sathiaraj D, Huang X, and Chen J. Predicting climate types for the Continental United States using unsupervised clustering techniques. *Environmetrics*. 2019;30(4):e2524.

[15]   Lizundia-Loiola J, Otón G, Ramo R, *et al*. A spatio-temporal active-fire clustering approach for global burned area mapping at 250 m from MODIS data. *Remote Sensing of Environment*. 2020;236:111493.

[16] Sharma B, Kumar M, Denis DM, *et al.* Appraisal of river water quality using open-access earth observation data set: a study of river Ganga at Allahabad (India). *Sustainable Water Resources Management.* 2019;5(2):755–765.

[17] Firozjaei MK, Sedighi A, Argany M, *et al.* A geographical direction-based approach for capturing the local variation of urban expansion in the application of CA-Markov model. *Cities.* 2019;93:120–135.

[18] Wang L, Dong Q, Yang L, *et al.* Crop classification based on a novel feature filtering and enhancement method. *Remote Sensing.* 2019;11(4):455.

[19] Foody GM, Ling F, Boyd DS, *et al.* Earth observation and machine learning to meet Sustainable Development Goal 8.7: mapping sites associated with slavery from space. *Remote Sensing.* 2019;11(3):266.

[20] Newcomers Earth Observation Guide. Available from: https://business.esa.int/newcomers-earth-observation-guide.

[21] Earth observation: Data, processing and applications – CRC for spatial ... Australia and New Zealand CRC for Spatial Information. Available from: https://www.crcsi.com.au/assets/Consultancy-Reports-and-Case-Studies/Earth-Observation-reports-updated-Feb-2019/Vol1-Introduction-low-res-0.8MB.pdf.

[22] Vasefi F, MacKinnon N, and Farkas D. Hyperspectral and multispectral imaging in dermatology. In: *Imaging in Dermatology.* New York, NY: Elsevier; 2016. p. 187–201.

[23] Principles of Remote Sensing and Processing. Available from: https://crisp.nus.edu.sg/ research/tutorial/opt_int.htm.

[24] Satellite imagery and archive, 2021. Available from: https://www.planet.com/products/planet-imagery/.

[25] Wei L, Wang K, Lu Q, *et al.* Crops fine classification in airborne hyperspectral imagery based on multi-feature fusion and deep learning. *Remote Sensing.* 2021;13(15):2917.

[26] What is Thermal Infrared Imaging, 2022. Available from: https://movitherm.com/knowledgebase/thermal-infrared-imaging.

[27] Teena M and Manickavasagan A. Thermal infrared imaging. In: *Imaging with Electromagnetic Spectrum.* New York, NY: Springer; 2014. p. 147–173.

[28] OpenSnow. Infrared Satellite – Explained. Available from: https://opensnow.com/news/post/infrared-satellite-explained.

[29] Rap Real-Time Weather. Available from: https://weather.rap.ucar.edu/.

[30] Agarwal T. Radar – Basics, Types, Working, Range Equation & Its Applications, 2020. Available from: https://www.elprocus.com/radar-basics-types-and-applications/.

[31] Radar Course 2. Available from: https://earth.esa.int/eogateway/missions/ers/radar-courses/radar-course-2#: :text=Real

[32] Remote Sensing GIS Applications 3(2+1). Available from: http://ecourse-sonline.iasri.res.in/mod/page/view.php?id=2068.

[33] Sentinel Datset. Available from: https://scihub.copernicus.eu/dhus/#/home.

[34] Narain A. Data Analytics to Shape the Future of Earth Observation, 2019. Available from: https://www.geospatialworld.net/blogs/future-of-earth-observation/.

[35]   Pesaresi M, Syrris V, and Julea A. A new method for earth observation data analytics based on symbolic machine learning. *Remote Sensing.* 2016;8 (5):399.

[36]   Yang C, Yu M, Li Y, *et al.* Big Earth data analytics: a survey. *Big Earth Data.* 2019;3(2):83–107.

[37]   Big Earth Data Observation. Available from: https://ati.ec.europa.eu/sites/ default/files/2020-06/Big%20Data

[38]   Haase D, Jänicke C, and Wellmann T. Front and back yard green analysis with subpixel vegetation fractions from earth observation data in a city. *Landscape and Urban Planning.* 2019;182:44–54.

[39]   Dogan T and Uysal KA. The impact of feature selection on urban land cover classification. *International Journal of Intelligent Systems and Applications in Engineering.* 2018;6:59–64.

[40]   Marmanis D, Datcu M, Esch T, *et al.* Deep learning earth observation classification using ImageNet pretrained networks. *IEEE Geoscience and Remote Sensing Letters.* 2015;13(1):105–109.

[41]   Yi Y and Zhang W. A new deep-learning-based approach for earthquake-triggered landslide detection from single-temporal RapidEye satellite imagery. *IEEE Journal of Selected Topics in Applied Earth Observations and Remote Sensing.* 2020;13:6166–6176.

[42]   Hoeser T and Kuenzer C. Object detection and image segmentation with deep learning on earth observation data: a review – Part I: evolution and recent trends. *Remote Sensing.* 2020;12(10):1667.

[43]   Som-ard J, Immitzer M, Vuolo F, *et al.* Mapping of crop types in 1989, 1999, 2009 and 2019 to assess major land cover trends of the Udon Thani Province, Thailand. *Computers and Electronics in Agriculture.* 2022;198:107083.

[44]   Kuchler PC, Simões M, Ferraz R, *et al.* Monitoring complex integrated crop–livestock systems at regional scale in Brazil: a big Earth Observation data approach. *Remote Sensing.* 2022;14(7):1648.

[45]   Memon N, Parikh H, Patel SB, *et al.* Automatic land cover classification of multi-resolution dualpol data using convolutional neural network (CNN). *Remote Sensing Applications: Society and Environment.* 2021;22:100491.

[46]   Löw F, Stieglitz K, and Diemar O. Terrestrial oil spill mapping using satellite earth observation and machine learning: a case study in South Sudan. *Journal of Environmental Management.* 2021;298:113424.

[47]   Seydi ST, Hasanlou M, Amani M, *et al.* Oil spill detection based on multi-scale multidimensional residual CNN for optical remote sensing imagery. *IEEE Journal of Selected Topics in Applied Earth Observations and Remote Sensing.* 2021;14:10941–10952.

[48]   Hong Z, Yang T, Tong X, *et al.* Multi-scale ship detection from SAR and optical imagery via a more accurate YOLOv3. *IEEE Journal of Selected Topics in Applied Earth Observations and Remote Sensing.* 2021;14:6083–6101.

[49]   El Naqa I and Murphy MJ. What is machine learning? In: *Machine Learning in Radiation Oncology.* New York, NY: Springer; 2015. p. 3–11.

[50] Qiu J, Wu Q, Ding G, *et al.* A survey of machine learning for big data processing. *EURASIP Journal on Advances in Signal Processing.* 2016;2016 (1):1–16.

[51] Picoli MCA, Camara G, Sanches I, *et al.* Big earth observation time series analysis for monitoring Brazilian agriculture. *ISPRS Journal of Photogrammetry and Remote Sensing.* 2018;145:328–339.

[52] Löw F, Michel U, Dech S, *et al.* Impact of feature selection on the accuracy and spatial uncertainty of per-field crop classification using support vector machines. *ISPRS Journal of Photogrammetry and Remote Sensing.* 2013;85:102–119.

[53] Foerster S, Kaden K, Foerster M, *et al.* Crop type mapping using spectral–temporal profiles and phenological information. *Computers and Electronics in Agriculture.* 2012;89:30–40.

[54] Collobert R and Weston J. A unified architecture for natural language processing: deep neural networks with multitask learning. In: *Proceedings of the 25th International Conference on Machine Learning,* 2008. p. 160–167.

[55] Krizhevsky A, Sutskever I, and Hinton GE. Imagenet classification with deep convolutional neural networks. *Communications of the ACM.* 2017;60 (6):84–90.

[56] Schmidhuber J. Deep learning in neural networks: an overview. *Neural Networks.* 2015;61:85–117.

[57] Zhong L, Hu L, and Zhou H. Deep learning based multi-temporal crop classification. *Remote Sensing of Environment.* 2019;221:430–443.

[58] Hoeser T, Bachofer F, and Kuenzer C. Object detection and image segmentation with deep learning on Earth observation data: a review. Part II: applications. *Remote Sensing.* 2020;12(18):3053.

[59] Goyal A. Must known data visualization techniques for Data Science. *Analytics Vidhya.* 2021; Available from: https://www.analyticsvidhya.com/blog/2021/06/must-known-data-visualization-techniques-for-data-science/.

[60] Lesley. 7 techniques to visualize geospatial data – Atlan: Humans of data, 2019. Available from: https://humansofdata.atlan.com/2016/10/7-techniques-to-visualize-geospatial-data/.

[61] 12 Methods for Visualizing Geospatial Data on a Map. Available from: https://www.safegraph.com/guides/visualizing-geospatial-data.

[62] GISGeography. Cartogram Maps: Data Visualization with Exaggeration, 2022. Available from: https://gisgeography.com/cartogram-maps/.

[63] Geothermal Heat Map. Available from: https://serc.carleton.edu/details/images/46195.html.

[64] Rice S. Healthcare Scorecard Shows Little Progress Among U.S. States. *Modern Healthcare.* 2014; Available from: https://www.modernhealthcare.com/article/20140430/NEWS/304309964/healthcare-scorecard-shows-little-progress-among-u-s-states.

[65] Space Applications Center, Indian Space Research Organization, Government of India. Available from: https://vedas.sac.gov.in/en/.

[66]   Applications of Earth Observation. Available from: https://www.ukspace. org/wp-content/uploads/2019/05/The-many-uses-of-Earth-Observation-data. pdf.

[67]   Hydrosphere. Available from: https://www.sac.gov.in/Vyom/hydrosphere. jsp.

[68]   Wetlands.   Available   from: https://www.isro.gov.in/earth-observation/ wetlands.

Part I

Clustering and classification of Earth Observation data

Chapter 2

Deep learning method for crop classification using remote sensing data

Kavita Bhosle[1] and Vijaya Musande[2]

The Red Edge Position (REP) index plays an important role in agricultural remote sensing applications. A wavelength from 350 nm to 990 nm is the common range for green. In this chapter, we have focused mainly on crop classification using the deep learning method. We have presented a study of crop classification using deep learning methods on hyperspectral remote sensing data. Deep learning is the evolved form of artificial neural network (ANN). It is based on a biological concept that deals with the network of neurons in a brain. To solve problems regarding crop classification, many machine-learning methods are used by researchers. Traditional machine-learning algorithms, including support vector machine, decision tree-based, and random forest, work on structured data only. Remote sensing data is unstructured data. Hence more computational overheads are needed to organize the unstructured data into structured ones. One of the most adaptable state-of-the-art approaches for feature extraction and classification of unstructured and structured data is deep learning. Thus we have focused on deep learning convolutional neural network (CNN) for feature extraction and classification of crops.

2.1 Sources of remote sensing data collection

There are multiple sources available for the collection of remote sensing data. Sensors are mounted on either satellites or aircraft. Each data has its own spectral and spatial resolution. Spectral resolution defines intervals between two consecutive wavelengths. The finer the spectral resolution, the narrower the wavelength range for a particular channel or band. There are two types of data collection techniques used: active and passive. In passive, reflected sunlight is used to measure radiation. These types of sensors can work in the presence of sunlight. Active sensors are not dependent on the Sun's electromagnetic rays. These sensors use their electromagnetic energy and incident on the earth's surface. Reflected energy is collected by active sensors.

[1]MIT, India
[2]Jawaharlal Nehru Engineering College, MGM University, India

Spectral reflectance can also be acquired with an ASD Spectro radiometer that provides measurements in the spectral range starting from 350 nm to 2,500 nm with 3 nm spectral resolution and a 1 nm sampling step. These experiments can be carried out in the field or laboratory. Obtained data can be viewed and exported in ASCII file using View Spec Pro 6.2.

2.2   Tools for processing remote sensing data

Quantum GIS (QGIS) is available at https://qgis.org/en/site/forusers/download.html, System for Automated Geoscientific Analyses (SAGA) is available at https://saga-gis. sourceforge.io/en/index.html, Geographic Resources Analysis Support System (GRASS) is available at https://grass.osgeo.org/, The Integrated Land and Water Information System (ILWIS) is available at https://www.itc.nl/ilwis/, etc. are the tools available for processing remote sensing data. Radiometric, geometry and atmospheric correction is required in GIS. In the process of remote sensing, the atmosphere gives radiation and is mixed with earth radiation. Atmospheric correction is required to remove bands created by atmospheric error. Hyperspectral data has many numbers of bands or dimensions. Selection techniques can be used for dimensionality reduction [1]. Principal component analysis (PCA) and linear discriminant analysis (LDA) are transformation techniques that play a significant role in the dimensionality reduction of hyperspectral images [2]. Strong spectral correlation in hyperspectral photographs results in redundant information. Therefore, it is necessary to minimise these photographs' dimensions. Techniques for selection or transformation can be used to reduce dimensionality. PCA is a transformation method that is important for reducing the dimensionality of hyperspectral pictures. From hundreds of hyperspectral bands, interesting bands have been extracted using kernel PCA.

Hyperspectral data requires dimensionality reduction since it contains a large number of bands. Prior to applying a machine-learning algorithm to the data, the dimensionality-reduction approach PCA has been used to convert a high-dimensional dataset into a smaller-dimensional subspace. There is a tonne of information in the thousands of bands of hyperspectral data. Using the Eigenvalues and the Eigenvectors, PCA generates valuable and significant principal components.

As a segment for the experiment, the region of interest (ROI) has been extracted. For ROI, there is empirical data. In order to obtain useful components and reduce dimensionality, PCA has been performed. Using these elements and ground truth, patches are produced. After that, the training and testing datasets are created from these patches. This data has been subjected to a deep-learning CNN model. Various CNN parameters have been changed, and the model has been assessed using precision, recall, f1 score and test accuracy.

2.3   Crop classification using remote sensing data

Monospectral, multispectral, hyperspectral and many more types of remote sensing data are available. Monospectral has only one band, whereas multispectral has

10–12 bands which are not contiguous [3]. Hyperspectral data has hundreds of contiguous spectral bands. Hyperspectral imaging is the collection of hundreds of continuous connecting spectral bands which can be employed to represent each pixel [4].

2.3.1 Methods for crop classification

Conventional methods for crop classification are neural networks, random forest method, support vector machines and many more [5]. Deep learning is the evolved form of the ANN [6]. It is based on a biological concept that deals with the network of neurons in a brain. To solve problems regarding crop classification, many machine-learning methods are used by researchers. Traditional machine-learning algorithms, including support vector machine, decision tree-based, and random forest, work on structured data only. Remote sensing data is unstructured data [7]. Hence, more computational overheads are needed to organize the unstructured data into structured ones. One of the most adaptable states-of-the-art approaches for feature extraction and classification of unstructured and structured data is deep learning [8]. Thus, we have presented a study using deep learning CNN for feature extraction and classification of crops [9]. EO-1 hyperspectral images provide an appropriate spectral and spatial resolution. This helps to classify the crops much better than the conventional methods. EO-1 Hyperion data is available at https://www.usgs.gov/centers/eros/science/earth-observing-1-eo-1. EO-1 Hyperion has given 242 bands data of 10 nm spectral resolution and 30 m spatial resolution. Only 155 bands are received after performing atmospheric correction. The study area of (49 × 69) consists of 3,381 pixels [10]. The advanced version of an ANN is deep learning. It is based on a biological idea that pertains to the brain's network of neurons. Numerous machine-learning techniques are employed by researchers to address crop classification issues. Support vector machines, decision trees and random forests are examples of traditional machine-learning algorithms that only function with structured input. Data from remote sensing is unstructured. Therefore, converting unstructured data to structured data requires higher computing overhead. Deep learning is one of the most flexible modern methods for feature extraction and categorization of unstructured and structured data. We, therefore, concentrated on deep learning CNN for feature extraction and crop categorization.

The REP index plays an important role in agricultural remote sensing applications. A wavelength from 350 nm to 990 nm is the common range for green. REP is different for each crop [11,12]. This value can be used for discrimination of crops as well as conditions of crops. Crop conditions may be healthy or diseased crops [13]. The REP index can be calculated using many methods like linear extrapolation method, maximum first derivative and language interpolation. REP value can be used for crop classification [14,15].

Experiments to assess plant health have been conducted using healthy and diseased samples of mulberry, cotton and sugarcane plants. For locating REP, four techniques have been investigated. The simplest method is the maximum derivative. The easiest way to determine REP is linear interpolation, which takes into account both the maximum and lowest shoulders of the chlorophyll reflectance curve. For all crops and environmental conditions, the third linear extrapolation has taken into account the fixed four spots on the chlorophyll reflectance curve. The two approaches mentioned above are typically utilised when there are two peaks in the first derivative and a nitrogen-affected REP.

A narrowband reflectance capability that responds to variations in chlorophyll content is called the REP index. Increasing chlorophyll concentration broadens the absorption characteristic and shifts the red edge to longer wavelengths. Red edge location refers to the wavelength with the steepest slope between 690 and 740 nm. Green vegetation typically has a wavelength between 700 and 730 nm.

2.3.2   Case study

Space-borne hyperspectral remote sensing data (EO-1 Hyperion) dataset has been used in the present study. This data consists of 0m spatial and 10 nm spectral resolution. The study area is selected from the Aurangabad District region in Maharashtra, India. Data have gathered by conducting a campaign in the month of December 2015. The selected region is segmented using the ENVI tool. This segment of ROI consists of fields of sugarcane, cotton, and mulberry plantations and consists of rock as shown in Figure 2.1. This figure has been referred from the research paper [10].

Labelling of ROI data corresponding to individual pixels which have been labelled using collected ground truth. The final labelled training and testing dataset has been formed using this process. Deep learning CNN has been used for feature extraction and classification [16]. CNN consists of alternate convolutional and max pooling layer followed by a fully connected neural network [10] as shown in Figure 2.2.

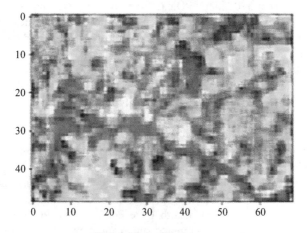

Figure 2.1   Study area

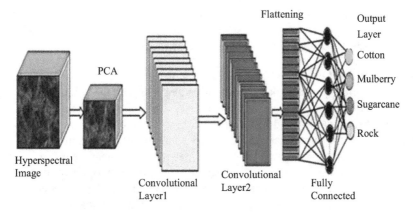

Figure 2.2   Architecture of CNN

Data have been gathered via campaigns. The chosen terrain includes of rock as well as sugarcane, cotton, mulberry and a limited quantity of other crop fields with different plant heights between 0.7 and 1.5 m. Using gathered ground truth data as well as the latitude and longitude of each pixel, the labelling of hyperspectral data corresponding to individual pixels has been completed. This method resulted in the creation of the final labelled training and testing dataset. Experts who obtained data in the field manually marked various crop varieties. The extracted and labelled hyperspectral data corresponding to each pixel were then created using the combined ground truth data. The final labelled dataset used for training and testing was created through this approach. The correctly labelled dataset for crop classification was created by running a data-gathering effort.

Python experiments have been carried out utilising the Keras library functions and the TensorFlow environment.

2.3.2.1 Convolutional layers

Consider d × d input hyperspectral segment given input to convolutional layer. We have given p × p filter w; the output of the convolutional layer will be of the size $(d - p + 1) \times (d - p + 1)$ [17,18]. In (2.1), after summing up the contributions, weighted by the filter components from the previous layer cells, adding bias term and then applying the activation function [19].

Using the softmax function, it has been connected to a dense layer that is entirely connected. The activation functions of an ANN determine whether or not a neuron should fire and whether the information it contains is meaningful. The hidden layers have been implemented using rectified linear units (ReLUs). A rectified linear unit outputs × when × is positive and 0 when × is negative. CNN only has one channel and uses 2 × 2 and 3 × 3 filter sizes. Because more than two classes or crops need to be detected, the CNN output layer uses the softmax function. Each unit's corresponding output, between 0 and 1, has been provided.

Following are the steps that CNN took to train the model: The activation functions ReLU and 3 × 3 filter size have been employed, and the learning rate has been held constant at 0.0001. The batch size has been held constant at 32. CNN has two convolutional layers in its proposed design, which are followed by fully connected layers and output layers, as shown in Figure 2.2.

2.3.2.2   Max pooling layers

The max-pooling layers are k × k box and output a single value which is optimized for that section. For example, if the input layer is d × d layer, then the output will be computed as a (d/k) × (d/k) layer, k × k block is compact to presently a single value through the max function. ReLu and Tanh are the activation function [20]. The output of the max pooling layer has to flatten and then can be given to a fully connected layer with an activation function [21]. The softmax function is used in this case in the output layer as class labels are not binary. In this case study, output layer consists of four nodes of classes, mulberry, cotton, sugarcane and rocks.

2.3.2.3   Convolutional autoencoder

Stack autoencoder and convolutional autoencoder are similar terms (SAE). In many unsupervised applications, dimensionality reduction is primarily achieved via autoencoders and neural networks. Additionally, they are employed in image retouching, image enhancement and image colourisation. Deep learning feature extraction has been carried out using a multiscale convolutional autoencoder. To reduce noise and improve the image in this investigation, we used a convolutional autoencoder. It involves a two-step technique. Encoding and decoding are the first and second steps, respectively. The input is transformed through encoding into a latent state space with fewer dimensions. Utilising upsampling techniques, the decoding process attempts to reconstitute the input. The input of size $n \times n$ is transformed into $(n - m + 1) \times (n - m + 1)$ by the convolutional layer.

Latent state space is minimised by using numerous convolutional layers autoencoders. The decoder then employs a convolutional layer and several upsampling layers to obtain the image's original size. To produce a magnified image, upsampling repeats the data's rows and columns by size [0] and size [1], respectively. CNN models have been contrasted with deep neural network (DNN) models and convolutional autoencoder models.

2.3.2.4   Deep neural network

A multilayered feed-forward perceptron, which consists of multiple layers of neurons coupled to one another, is one type of neural network that is frequently employed. In several sectors, remote sensing has been successfully implemented using ANNs. Deep neural networks are multilayer perceptron (MLP) systems with deep hidden layers. The complexity of a deep neural network rises with the number of hidden layers.

In this study, there are four dense layers that make up deep NN. Thirty nodes and a ReLU activation function are present in each layer. The Deep NN classifier can be thought of as a parallel computing system made up of a very large number of basic inter connected processors.

2.4 Performance evaluation

Deep learning CNN model has been implemented on this data. The model has been evaluated using precision, recall, F1 score and test accuracy. About 75% of the samples have been used for training, and 25% have been used for testing. CNN has been compared with two different supervised classification methods, deep neural network and convolutional autoencoder [22]. It has been observed that Adam, with ReLU activation function and filter size of 2×2, has given better classification accuracy compared to the other methods [23,24]. About 70% data has been used for training the model, and 30% data has been used for testing purposes.

PCA and CNN experiments are carried out on the Python Tensorflow environment. The PCA receives as its input unstructured hyperspectral pictures. As these major components have provided more information, the experiment will only take into account the first 30 components.

Effective implementation of the predictive models has been possible for crop classification using deep learning. Deep learning CNN extracts features and then classified the crops [25–27]. As shown in Table 2.1, the overall accuracy of CNN is 88 ± 2.43; convolutional autoencoder is 85 ± 1.57. DNN has given 78 ± 1.15 accuracy. REP also plays an important role in accuracy [28].

Table 2.2 shows the confusion matrix for the CNN model for four classes; Cotton, Mulberry, Sugarcane and Rock have a number of samples as 269, 113, 128 and 345, respectively. Overall accuracy is calculated using a confusion matrix. The confusion matrix shows actual and predicted values for four classes. Using the following formulae, precision, recall, F1 score and overall accuracy have been calculated. It has been observed that accuracy using the CNN method is good compared to convolutional autoencoder and deep NN method. As there are more

Table 2.1 Overall accuracy comparison of CNN with other methods

	% Overall accuracy	
CNN	Convolutional autoencoder	Deep NN
88 ± 2.43	85 ± 1.57	78 ± 1.15

Table 2.2 Confusion matrix for CNN method

		Predicted values			
Actual values		0 (Cotton)	1 (Mulberry)	2 (Sugarcane)	3 (Rock)
	0 (Cotton)	TP 0	E 01	E 02	E 03
	1 (Mulberry)	E 10	TP 1	E 12	E 13
	2 (Sugarcane)	E 20	E 21	TP 2	E 23
	3 (Rock)	E 30	E 31	E 32	TP 3

continuous spectral bands in the hyperspectral dataset, the signature of each crop is more obvious. This method of crop classification works well. Because it effectively handles unstructured data, deep-learning CNN has been chosen. It can automatically extract the information needed for crop detection or categorization. To obtain an optimum CNN model, a fine-tuning strategy has been applied by adjusting the values of various parameters.

DNNs and convolutional autoencoder, two distinct supervised classification techniques, are contrasted with optimised CNN. The model's performance can be enhanced by employing precisely calibrated parameters. Data for training and testing have been created through the use of significant field research. The samples were taken using a random sampling strategy. In Aurangabad, cotton is regarded as a significant winter crop. Sugarcane, mulberries and a few other minor crops are additional important crops. There are rocks all across the study area as well:

Precision 0 = (TP 0) / (TP 0 + E 10 + E 20 + E 30)
Recall 0 = (TP 0) / (TP 0 + E 01 + E 02 + E 03)
F1 Score = (2 * precision * recall) / (precision + recall)

Overall accuracy = number of correctly classified samples/number of test samples
Overall accuracy = (TP 0 + TP 1 + TP 2 + TP 3)/(total test record)

Three techniques: convolutional autoencoder, DNN, and optimized CNN have been used and compared. The improved CNN method has been found to produce superior classification outcomes compared to the other approaches. Through the use of flattened one-dimensional patches, convolutional autoencoders and DNNs can also extract deep features via deep learning, albeit the performance may be significantly lowered. Unsupervised learning is used by the convolutional autoencoder to acquire feature knowledge. As a result, it cannot properly process the information on the label.

Our goal was to create a crop classification system that had already been trained. This study concentrated on the efficient application of crop forecast predictive models. It also looks at how well the various predictive algorithms can classify data. The overall classification accuracy on the dataset can be significantly increased by the optimised CNN. It also functions with a minimal number of training samples, as was seen with the dataset for the research area.

The current study may be viewed as the initial stage in creating a crop categorization model that has already been trained. Additionally, it demonstrates that CNN's use of the Adam optimiser enhances efficiency even with sparse input. To improve accuracy on other crops, it is possible to build and test deep learning-based systems for remote sensing data classification in the future.

The signature of each crop is more distinct in the hyperspectral data set since it has more bands. Crop identification can be aided by this hyperspectral data feature. Typically, the confusion matrix is used to test classification accuracy. The outcomes of crop classification are compared to and examined using the confusion matrix, which is made up of the actual pixel classification results. The proportion of correctly classified pixels to all pixels is used to measure classification accuracy overall. A fully connected neural network layer follows the encoder-decoder layers of the convolutional autoencoder. Dimensionality is reduced during the encoding and decoding process, and it is possible that less valuable information is preserved as well. As a result, its accuracy is lower than CNN's.

In order to extract features, the convolution layer with an activation function reduces the size of the input image. It is accomplished by applying a weighted area filter on the input data. The data with the highest value within a region is accepted by the pooling layer. The function of these layers is to choose an essential attribute from the input for categorization. DNNs are unable to achieve this since they have several layers with a single activation function. Consequently, it is not possible to extract the particular feature. A fully connected neural network and an encoder-decoder are combined to create a convolutional autoencoder. Its accuracy therefore falls between that of the CNN and the deep neural network.

2.5 Conclusion

Hyperspectral remote sensing data has a greater number of bands; therefore, it has also been essential to reduce dimensions. The overall classification accuracy with a small number of training samples can remarkably improve by the CNN model. Each crop has a more prominent signature because the hyperspectral data set has a greater number of bands. Such type of research improves the perception of crop classification and its role in an agricultural field. It will help the farming community to plan suitable crops and farming practices and that will increase yield. Extension of this work can be used for finding crop sown area and expected yield. It also can be implemented for classifying and distinguishing other objects. Instead of only hyperspectral imaging, we can use multisource data that is a combination of microwave, hyperspectral and ultra-spectral sensors.

Healthy and diseased crops can monitor using REP and can be classified using deep learning methods. Deep learning methods can work on unstructured data like images, video and audio efficiently. It has been observed that Adam, with ReLU activation function and a filter size of 2×2, has given better classification accuracy compared to the other methods.

A pre-trained architecture for crop classification was to be developed as part of this work. DNN and convolutional autoencoder have both been compared to optimised models. For crop classification, it has been feasible to implement the predictive models successfully. The ability of the various analytical models to identify things is also looked at.

The overall classification accuracy of the dataset is seen to be significantly improved by the optimised CNN; it has been noticed. The suggested CNN model performs admirably on small training samples, as was noted in the case of the study area dataset.

It has also been crucial to lowering dimensions because hyperspectral remote sensing data contains a greater number of bands. From 155 to 36 bands could be successfully reduced using PCA. The results of this study have demonstrated the importance of crop identification in farming and the research community.

The future focus of this research will be on estimating crop yield and area seeded, classifying and identifying other items, and using multisource data from a mix of microwave, hyper-spectral and ultra-spectral sensors.

References

[1]   M. Fauvel, J. Chanussot, and J.A. Benediktsson (2009). Kernel principal component analysis for the classification of hyperspectral remote sensing data over urban areas. *Journal on Advances in Signal Processing*, 2009, Article ID 783194, 14 pages, doi:10.1155/2009/783194.

[2]   W. Zhao and S. Du (2016). Spectral–spatial feature extraction for hyper-spectral image classification: a dimension reduction and deep learning approach. *IEEE Transactions on Geoscience and Remote Sensing*, 54 (8):4544–4554.

[3]   C. Yang, C.P.-C. Suh, and J.K. Westbrook (2017). Early identification of cotton fields using mosaicked aerial multispectral imagery. *Journal of Applied Remote Sensing*, 11(1):016008, doi:10.1117/1.JRS.11.016008.

[4]   S.E. Hosseini Aria, M. Menenti, and B.G. H. Gorte (2017). Spectral region identification versus individual channel selection in supervised dimension-ality reduction of hyperspectral image data. *Journal of Applied Remote Sensing*, 11(4):046010, doi:10.1117/1.JRS.11.046010.

[5]   A. Ozdarici Ok, O. Akar, and O. Gungor (2012). Evaluation of random forest method for agricultural crop classification. *European Journal of Remote Sensing*, 45(1):421–432, doi: 10.5721/EuJRS20124535.

[6]   A. Zhang, X. Yang, L. Jia, J. Ai, and Z. Dong (2019). SAR image classifi-cation using adaptive neighbourhood-based convolutional neural network. *European Journal of Remote Sensing*, 52(1):178–193, doi: 10.1080/22797254.2019.1579616.

[7]   T. Tian, L. Gao1, W. Song, K.-K. Raymond Choo, and J. He (2017). *Feature Extraction and Classification of VHR Images with Attribute Profiles and Convolutional Neural Networks*. Springer Multimed Tools Appl, https://doi.org/10.1007/s11042-017-5331-4.

[8]   S. Yu, S. Jia, and C. Xu (2017). Convolutional neural networks for hyper-spectral image classification. *Elsevier Neurocomputing*, 219:88–98.

[9]   Y. Chen, H. Jiang, C. Li, X. Jia, and P. Ghamisi (2016). Deep feature extraction and classification of hyperspectral images based on convolutional neural networks. *IEEE Transactions on Geoscience and Remote Sensing*, 54:6232–6251.

[10]   K. Bhosle and V. Musande (2019). Evaluation of deep learning CNN model for land use land cover classification and crop identification using hyper-spectral remote sensing images. *Journal of the Indian Society of Remote*

Sensing, 47(11):1949–1958, https://doi.org/10.1007/s12524-019-01041-2 (0123456789).

[11] K. Bhosle and V. Musande (2017). Stress monitoring of mulberry plants by finding rep using hyperspectral data. *The International Archives of Photogrammetry, Remote Sensing and Spatial Information Sciences*, 42:383.

[12] T. P. Dawson and P. J. Curran (1998). A new technique for interpolating the reflectance red edge position. *International Journal of Remote Sensing*, 19 (11):2133–2139.

[13] P. J. Zarco-Tejada, J. R. Miller, D. Haboudane, N. Tremblay, and S. Apostol (2003). Detection of chlorophyll fluorescence in vegetation from airborne hyperspectral CASI imagery in the red edge spectral region. In: *2003 IEEE International Geoscience and Remote Sensing Symposium. Proceedings* (IEEE Cat. No. 03CH37477), IEEE, 0-7803-7929-2/03.

[14] A.A. Gitelson, M.N. Merzlyak, and H.K. Lichtenthler (1996). Detection of red edge position and chlorophyll content by reflectance measurements near 700 nm. *Journal of Plant Physiology*, 148:501–508.

[15] A.A. Gitelson, M.N. Merzilyak, and H.K. Lichtenthaler (1996). Detection of red edge position and chlorophyll content by reflectance measurements near 700 nm. *Journal of Plant Physiology*, 148:501–508.

[16] A.A. Gitelson, M.N. Merzilyak, and H.K. Lichtenthaler (1996). Detection of red edge position and chlorophyll content by reflectance measurements near 700 nm, *Journal of Plant Physiology* 148:501–508.

[17] K. Bhosle and V. Musande (2020), Evaluation of CNN model by comparing with convolutional autoencoder and deep neural network for crop classification on hyperspectral imagery. *Geocarto International*, 37:813–827, doi: 10.1080/10106049.2020.1740950

[18] Y. Zhong, F. Fei, and L. Zhang (2016). Large patch convolutional neural networks for the scene classification of high spatial resolution imagery. *Journal of Applied Remote Sensing*, 10(2):025006, doi:10.1117/1. JRS.10.025006.

[19] Y. Li, H. Zhang, and Q. Shen (2017). Spectral–spatial classification of hyperspectral imagery with 3D convolutional neural network. *Remote Sensing*, 9:67, doi:10.3390/rs9010067.

[20] X. Yu and H. Dong (2018). *PTL-CFS Based Deep Convolutional Neural Network Model for Remote Sensing Classification*. Springer Computing, https://doi.org/10.1007/s00607- 018–0609-6.

[21] B. Jin, P. Ye, X. Zhang, W. Song, and S. Li (2019). Object-oriented method combined with deep convolutional neural networks for land-use-type classification of remote sensing images. *Journal of the Indian Society of Remote Sensing*, 47:951–965, https://doi.org/10.1007/s12524-019-00945-3 (0123456789).

[22] Z. Sun, X. Zhao, M. Wu, and C. Wang (2018). Extracting urban impervious surface from WorldView-2 and airborne LiDAR data using 3D convolutional neural networks. *Journal of the Indian Society of Remote Sensing*, 47:401–412, https://doi.org/10.1007/s12524-018-0917-5.

[23] N. Kussul, M. Lavreniuk, S. Skakun, and A. Shelestov (2017). Deep learning classification of land cover and crop types using remote sensing data. *IEEE Geoscience and Remote Sensing Letters*, 14(5):778–782.

[24] T. Fu, L. Ma, M. Li, and B.A. Johnson (2018). Using convolutional neural network to identify irregular segmentation objects from very high-resolution remote sensing imagery. *Journal of Applied Remote Sensing*, 12(2):025010, doi: 10.1117/1.JRS.12.025010.

[25] C. Zhang, J. Liu, F. Yu, *et al.* (2018). Segmentation model based on convolutional neural networks for extracting vegetation from Gaofen-2 images. *Journal of Applied Remote Sensing*, 12(4):042804, doi:10.1117/1. JRS.12.042804.

[26] A. Zhang, X. Yang, L. Jia, J. Ai, and Z. Dong (2019). SAR image classification using adaptive neighborhood-based convolutional neural network. *European Journal of Remote Sensing*, 52(1):178–193, doi: 10.1080/22797254.2019.1579616.

[27] Q. Yue and C. Ma (2016). Deep learning for hyperspectral data classification through exponential momentum deep convolution neural networks. *Hindawi Publishing Corporation Journal of Sensors,* 2016, Article ID 3150632, 8 pages, http://dx.doi.org/10.1155/2016/3150632.

[28] K.L. Smith, M.D. Steven, and J.J. Colls (2004). Use of hyperspectral derivative ratios in the red-edge region to identify plant stress responses to gas leaks. *Elsevier Remote Sensing of Environment*, 92:207–217.

Chapter 3

Using optical images to demarcate fields in L band SAR images for effective deep learning based crop classification and crop cover estimation

Kimee Joshi[1], Madhuri Bhavsar[1], Zunnun Narmawala[1] and Swati Jain[1]

This study aims at developing an end-to-end solution for deep learning-based crop classification using synthetic aperture radar (SAR) images. SAR provides all-weather, day, and night imaging capabilities, sensitive to dielectric properties. Optical images are intuitive and capture static information well, like the boundary of the field in the absence of atmospheric disturbances. In this work, the end-to-end solution to use deep learning algorithms for the crop-type classification is done using SAR images. The limitation of the SAR images is handled by using the boundary information from the optical data. For the classification of different crops in the test site, L-band ISRO L- & S-band Airborne SAR (ASAR) and Airborne Synthetic Aperture Radar (AIRSAR) images were acquired over an agricultural site near Bardoli and Flevoland respectively. Pre-trained model Inception v3 and Custom VGG like model were used for crop classification. Inception V3 enabled us to better discriminate crops, particularly banana and sugarcane, with 97% accuracy, while the Custom VGG like model achieved 95.17% accuracy for 11 classes.

3.1 Introduction

The contribution of agriculture to India's GDP (Gross Domestic Production) is 15.9% & India's 54.6% population is engaged in agriculture and allied activities. The COVID-19 pandemic and subsequent lockdowns have influenced most of the sectors of the economy. However, the agricultural sector has performed way better with a 2.9% growth rate amid 2019–2020, as against 2.74% accomplished amid 2018–2019 [1,2].

United Nations' Sustainable Development Goal is to eliminate global hunger, protect the indigenous seed and crop varieties, double agriculture productivity, and have small farmer incomes by 2030 [3]. To achieve this objective for successful

[1]Institute of Technology, Nirma University, India

production, it is necessary to calculate yield estimation and crop acreage information. This information is useful for making a national food policy [4]. Traditional methods for acquiring this information are very time and resource-consuming and require large manpower.

Remote sensing has performed a very important role to deal with a large-scale agriculture monitoring [5]. SAR imaging provides an effective solution for agriculture because it provides high-resolution, day and night imaging capability, weather independent, sensitive to dielectric properties such as biomass and water content. Various traditional classification techniques such as Support Vector Machine [6–8], Decision Tree [7], Maximum-Likelihood Classifier [9], Wishart Classifier [10], and Random Forest [11,12] have been applied on SAR images.

Deep learning-based classification methods have proven to perform better as compared to conventional classification methods. A significant number of excellent deep learning models have been used in the classification of optical images as well as SAR images in recent years. Deep learning automatically learns features from a dataset.

3.1.1 Motivation

Remote sensing has become a potent geospatial tool in agricultural systems. As it delivers valuable, precise, and timely information about agricultural urbanization at high spatial and spectral resolutions, SAR imaging is an effective agricultural solution. For crop classification, a variety of techniques were used, however, deep learning has shown to be more effective. As per the literature, many researchers have tried machine learning and deep learning algorithms for crop classification, but they have only worked on AIRSAR, Advanced Land Observation Satellite (ALOS), Polarimetric Phased Array L-band Synthetic Aperture Radar (PALSAR), Uninhabited Aerial Vehicle Synthetic Aperture Radar (UAVSAR), Experimental Synthetic Aperture Radar (ESAR), etc., satellite datasets. In contrast, we worked on the ASAR dataset (provided by SAC-ISRO), which will provide global coverage of the entire landmass. This paper aims to work on the above-mentioned dataset and compare the crop classification results with the existing dataset.

3.1.2 Research contribution

Following are the major research contributions of this paper:

- The paper starts with an extensive literature survey focusing on machine learning and deep learning-based techniques for crop classification.
- We demonstrate effective results using a deep learning-based edge detection technique for identifying the field boundaries from optical image.
- We used the boundary obtained in the optical image by superimposing it on the SAR image for the demarcation of fields in the SAR image.
- Extracted fields are then used using Inception v3 and Custom VGG like a model for crop classification.
- Performance evaluation of the proposed model has been done using the evaluation matrices such as user's accuracy, producer's accuracy, overall accuracy, and kappa coefficient.

3.1.3   Organization

This paper describes a deep learning model that does crop-type classification from full polarization L-band SAR images. The rest of the paper is organized as follows: Section 3.2 discusses the related work, Section 3.3 describes the methodology, Section 3.4 describes the study area, Section 3.5 shows an experimental setting, Section 3.6 presents the experimental result and analysis, and concluding remarks are in Section 3.7.

3.2   Related work

Following are the contributions of various researchers in this domain: most of the techniques are applied to the dual-polarization dataset. Various deep learning techniques are applied in the SAR domain such as autoencoder [4], convolutional neural networks (CNNs) [4,13], fully convolutional networks (FCN) [4,14], recurrent neural network (RNN) [5], long short-term memory (LSTM) RNNs [4,13], gated recurrent unit (GRU) RNN [4,13], CNN LSTM [15], BiConvLSTM [15], and ResNet (deep residual network) [16] for crop classification. Table 3.1 shows the comparative analysis of the existing machine learning and deep learning based crop classification schemes along with the proposed scheme.

References [17–21] have used a combination of multi-temporal SAR and optical data for crop classification. Jose *et al.* [17] have classified sub-tropical crop types using autoencoder and CNN on both SAR and optical image datasets. Laura *et al.* [4] have compared various approaches such as Random forest, Autoencoder, CNN, and FCN for crop classification. All the authors have concluded that the CNN model performs the best.

Gu *et al.* [18] have used CNN and VGG approaches for crop classification on both SAR and optical image datasets. Avolio *et al.* [19] have proposed optimal data imputation network (ODIN) for land cover and crop classification from SAR and multi-spectral data and have used various input bands such as raw data, calibrated data, and NDVI. Mei *et al.* [16] have used feature set optimization and super-pixel segmentation for crop classification. Mullissa *et al.* [14] have classified the crops from multi-temporal SAR images using FCN. Wei *et al.* [22] have used the U-Net model to classify large-scale crop mapping from multi-temporal SAR images. Chen *et al.* [23,24] have used a combination of CNN with Coherency matrix T3 elements and other polarimetric features for Multi-temporal data. Lingjja *et al.* [18] have analyzed that multi-source and multi-temporal data provide good crop classification accuracy. Sun *et al.* [5], Emile *et al.* [25], and Zhao *et al.* [13] have used a Recurrent Neural Network (RNN)-based approach for crop classification.

3.3   Proposed methodology

The proposed methodology is divided into three steps: (1) SAR image pre-processing and decomposition, (2) edge detection & field extraction, and (3) crop classification using deep learning. Figure 3.1 illustrates the processing steps for the methodology.

Table 3.1 Comparative analysis of existing ML and DL-based crop classification schemes

Author	Year	Objectives	Sensor	Frequency	Polarimetric parameters	Multi date	Area	No. of classes identified
Halder et al. [9]	2011	Evaluation L band data with different polarization combination for crop classification	ALOS PALSAR	L	Different polarization combination modes 1. Linear polarization 2. Circular polarization 3. Hybrid polarization		Hisar, Haryana	5
Henning [10]	2011	To study crop classification accuracy for different polarization modes	EMISAR	L & C	1. Single polarization 2. Coherent and incoherent dual polarizations 3. Fully polarimetric		Denmark	11
Ning et al. [26]	2016	Proposed an improved super pixel-based POLSAR image classification integrating color features	1. AIRSAR 2. ESAR	L	Pauli decomposition		1. Flevoland Dataset 1 2. Oberpfaffenhofen	1. 10 2. 2
Huapeng et al. [11]	2019	explored Capability of time series data for crop classification	UAVSAR	L	1. Cloude–Pottier decomposition 2. Freeman-Durden decomposition 3. Linear polarization (VV, HV, HH) 4. Combination of all		Sacramento Valley, California	11

Reference	Year	Objective	Sensor	Band	Features/Techniques	Study area	No.
Julien et al. [12]	2019	To identify the best SAR configuration to identify winter land use	RADARSAT-2 Sentinel ALOS-2	L & C	1. Back scattering coefficients 2. Cloude–Pottier decomposition 3. Freeman–Durden decomposition 4. SPAN (total scattered power) and Shannon entropy 5. Dual and quad 6. Polarization (pol) mod coherence matrix [T] 7. Decomposition Techniques	Mount saint Michel, France	12
Hongwei et al. [27]	2019	Proposed a differentiable architecture search (DAS) method	1. AIRSAR 2. AIRSAR 3. ESAR	L	1. Pauli 2. Cloude 3. Freeman 4. H/A/Alpha 5. Huynen 6. Yamaguchi 7. Krogager	1. San Francisco 2. Flevoland Dataset 1 3. Oberpfaffenhofen	1. 5 2. 15 3. 3
Chu et al. [28]	2020	Proposed a POLSAR image classification based on FCN and manifold graph embedding model	1. AIRSAR 2. AIRSAR 3. AIRSAR	L		1. Flevoland Dataset 1 2. San Francisco 3. Flevoland Dataset 2	11 5 14
Proposed approach	–	Proposed a deep learning based crop classification model	1. ASAR 2. AIRSAR	L	Freeman Decomposition	1. Bardoli 2. Flevoland	1) 2 2) 11

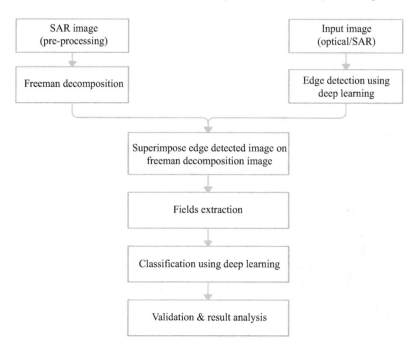

Figure 3.1   Proposed methodology

3.3.1   SAR image pre-processing and decomposition

SAR images are pre-processed according to the standard procedures which include speckle filtering, slant to ground range conversion, and Freeman decomposition. Freeman decomposition converts the co-variance matrix to three scattering components: surface scattering, double bounce scattering, and volume scattering [29].

3.3.2   Edge detection & field extraction

Edges are detected using the holistically-nested edge detection (HED) model. HED is an end-to-end deep learning-based edge detection method. The pre-trained model of HED is trained on the BSDS500Dataset [30]. It utilizes a trimmed VGG-like CNN which performs an image-to-image prediction task. This neural network acquires rich hierarchical edge maps which can detect the boundaries of crop fields from the image. This method also overcomes all the following limitations of the famous Canny Edge Detection method:

1. The lower and the upper values of the hysteresis thresholding must be manually set and hence it requires a lot of trials and error.
2. The values set for hysteresis thresholding for one image may not necessarily fit for other images having different illumination conditions.
3. Many pre-processing steps like noise removal, smoothening of images, grayscale conversion, and blurring were required before performing Canny Edge Detection [31].

The pre-trained HED model has been integrated with the deep neural network (DNN) module of OpenCV. Hence, the entire pipeline has been developed using OpenCV in Python. The pipeline consists of the following steps:

1. The Bilateral Filtering was applied to the input Optical/SAR image. This filtering method is known for the removal of noise in the image while preserving the edges. It compares the pixel intensity variation of the edge and the neighboring pixels of the edge which would be included for blurring [32].
2. The output image is then sent to the HED model which produces an edge map that does a good job conserving the boundaries of the crop fields. Hence, a binary image having the crop field boundaries is obtained.
3. The retrieval of contours is done on the binary image and every contour representing an individual crop field is extracted separately.

Segmentation of SAR image was done by super-imposing the edge detected image on Freeman decomposed image to extract out the fields. Crop fields for which ground truth data was available were labeled.

3.3.3 Classification using deep learning

Training a model from scratch required a large amount of data. So, pre-trained models which are trained using a large data set are used. The Custom VGG like model and the Inception v3 model is trained on the ImageNet Large Visual Recognition Challenge (ILSVRC) [33] dataset containing 1,000 classes containing over 1 million training images. For training purposes, the deep learning model requires a large amount of data. With less amount of data, the model will not achieve good accuracy over training as well as testing data. The existing dataset is so small because ground truth data is not available for all the fields. So, the existing dataset is augmented to make a better-generalized model. Data augmentation is the technique to increase the size of the dataset. Various techniques are used for augmentation such as flip, rotate, skew, and zoom. Figure 3.2 shows the various augmented images.

3.3.3.1 Inception V3

In recent years, Inception V3 model has been applied in many fields such as Human Eye Iris Recognition [34], maize disease detection in optical image [35], bird voice recognition [36], vehicle pollution detection [37], breast cancer

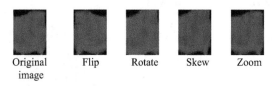

Original   Flip   Rotate   Skew   Zoom
image

Figure 3.2   Image augmentation

detection [38], actions of distracted drivers [39], skin disease detection [40], flower classification [41], and remote-sensing scene classification [42]. In Inception-V3, the number of connections is reduced without affecting the network efficiency using methods such as factorization of 5×5 convolutional layer into two 3×3 convolutional layers, factorization into asymmetric convolutions such as 3×3 into 1×3 and 3×1, and Auxiliary classifier. The model uses an input image size of 299*299 [43].

3.3.3.2   Custom VGG-like model

In Custom VGG like model, all the convolutional layers use filters of the size of 33 with stride=1, and for Max pooling, layers use a 2*2 filter size with 2 strides. The model uses an input image size of 64*64. With each convolutional layer, the number of filters doubles and with each pooling layer, the width and the height of the image reduces by half [44].

3.4   Study area

To validate the performance of the Inception V3 and Custom VGG like model, experiments are carried out on two datasets. The description of the two datasets is provided in Table 3.2.

Dataset 1: Bardoli area

The study area of Bardoli is located in the southern part of Gujarat state (see Figure 3.3). Bardoli is a suburb in Surat metropolitan region. It lies at 21.12°N 73.12°E [45]. The Bardoli area dataset obtained by Airborne SAR (provided by SAC-ISRO). The most common crops are Banana and Sugarcane.

Dataset 2: Flevoland area

The study area of Flevoland is located in the Netherlands (see Figure 3.4). Flevoland area data were obtained by the AIRSAR system of the NASA Jet Propulsion Laboratory in 1989 [46]. According to the Ground truth data provided in [28], the Flevoland area is divided into 11 types of different species: Bare Soil, Beet, Forest, Grass, Lucerne, Peas, Potato, Rapeseed, Steambean, Water and Wheat.

Table 3.2   Specifications of datasets used

Parameters	Dataset 1	Dataset 2
Sensor	ASAR	AIRSAR
Frequency	L-band	L-band
Data format	SLC	STK-MLC
Polarization	Full	Full
Test site	Bardoli, Gujarat	Flevoland, Netherlands
Acquisition date	17/06/2017	16/08/1989

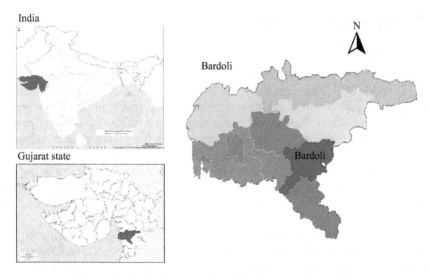

Figure 3.3   *Study area: Bardoli, Gujarat State, India*

Figure 3.4   *Study area: Flevoland, Netherlands*

3.5   Experimental setting

3.5.1   Dataset 1

In the experiment, SAR image was pre-processed according to the procedure such as Lee speckle filter, Freeman decomposition followed by conversion of slant range measurement into ground range measurement. The second is to detect the edges. So, edges were detected in optical image using a pre-trained Holistically-nested Edge Detection (HED) model. Optical image is taken from Google Earth using co-ordinates given in the metadata of the ASAR L band Airborne SAR image of the Bardoli (see Figure 3.5). Segmentation of the Freeman decomposed image is done by super-imposing the edge detected optical image on the Freeman decomposed gray scale image and extracting out the fields. A total of 1,083 fields were extracted. With the help of Ground Truth data, crops were labeled.

Initially, we have seven labeled images of the different crops (banana, sugarcane) and bare soil. Using various augmentation techniques, we increase the size of the dataset. To validate our results, we divide the training dataset into two subsets of training and testing. Tables 3.3 and 3.4 show the number of samples for training

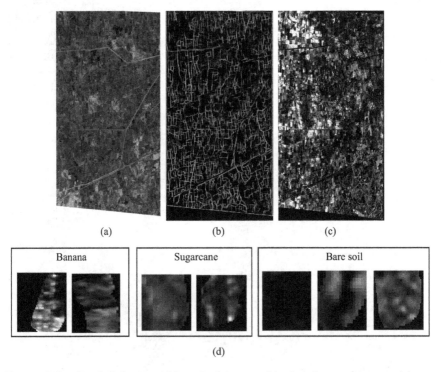

Figure 3.5:   Bardoli dataset: (a) optical image, (b) edge detected image, (c) Freeman decomposition image, and (d) samples of extracted fields mapped with GT

Table 3.3   Number of training and testing samples in dataset 1
after augmentation

Class name	Training samples	Testing samples
Cultivated area	2,002	202
Non-cultivated area	1,002	101
Total	3,004	303

Table 3.4   Number of training and testing samples in cultivated
area after augmentation

Class name	Training samples	Testing samples
Banana	1,001	101
Sugarcane	1,001	101
Total	2,002	202

and testing. The classification is done in two stages. In the first stage, Dataset 1 images are classified into two classes of cultivated area and non-cultivated area. In the second stage, cultivated areas are further classified into banana and sugarcane.

3.5.2   Dataset 2

Flevoland dataset was pre-processed according to the standard procedure such as Lee speckle filter, and Freeman decomposition (see Figure 3.6). Edges were detected in two different SAR images, i.e. Freeman decomposition image and the Pauli-decomposed image using the HED model. OR operation was performed on two edges detected images to generate a resultant image. The segmentation of the Freeman decomposition image (RGB image) is done by super-imposing the edge detected image (the resultant edge-detected image) on the Freeman decomposition image to extract out the fields. A total of 121 fields were extracted. Crops were labeled with the help of Ground Truth data provided in [28].

Initially, we have 127 labeled images of the 11 different classes: beet, lucerne, rapeseed, steambeans, peas, potato, wheat, bare soil, forest, grass, and water. Using the augmentation techniques, we increase the size of the dataset. To validate our results, we divide the training dataset into two subsets of training and testing. Table 3.5 shows the number of samples for training and testing after augmentation.

3.6   Experimental result and analysis

The training was performed using Inception V3 and Custom VGG like model. At each iteration, the accuracy and the loss of the model were recorded. The

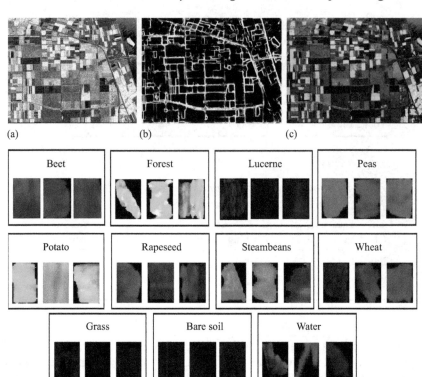

Figure 3.6   Flevoland dataset: (a) Pauli-decomposed image, (b) edge-detected image, (c) Freeman decomposition image, and (d) samples of extracted fields mapped with GT

Table 3.5   Number of training and testing samples in dataset 2 after augmentation

Class name	Training samples	Testing samples
Bare soil	1,508	101
Beet	1,529	103
Forest	1,502	101
Grass	1,519	102
Lucerne	1,504	101
Peas	1,505	101
Potato	1,537	103
Rapeseed	1,515	102
Steambeans	1,508	102
Water	1,504	101
Wheat	1,507	102
Total	16,638	1,119

performance of the model was assessed using cross-validation with the validation set. The training dataset was further classified into training and validation at the time of training. Inception v3 model achieved the best performance accuracy of 58.74% for Dataset 1 and 97.52% for the cultivated area. For Dataset 2, Custom VGG like model achieved the best performing accuracy of 95.17%.

The classification results of the proposed method are described from multiple perspectives. Confusion matrices for crop classification are shown in Figure 3.7 for Dataset 1, cultivated area (banana/sugarcane), and Dataset 2. Here it is seen that in the cultivated area, banana and sugarcane were often correctly classified, whereas in Dataset 1, while classifying cultivated and non-cultivated areas, cultivated area was misclassified. For Dataset 2, wheat is misclassified with rapeseed.

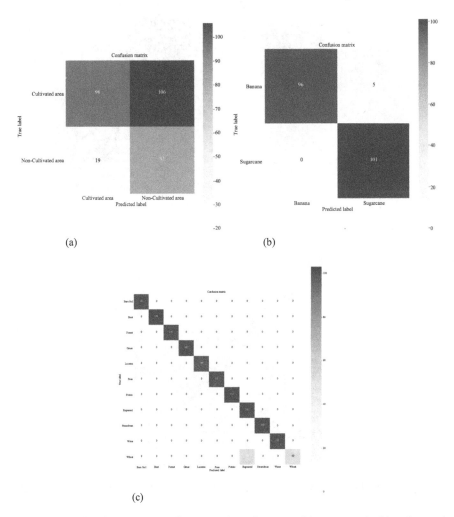

(a)   (b)

(c)

Figure 3.7   Confusion matrix for crop classification: (a) Dataset 1, (b) cultivated area, and (c) Dataset 2

The receiver operating characteristic (ROC) curve is plotted in Figure 3.8 for Dataset 1, cultivated area and Dataset 2. The cultivated area gives good accuracy over Dataset 1. Higher the ROC curve's area, better the model performance. In particular, the cultivated area was the best performing. For Dataset 2, all the values of curve are equal or near to 1 except wheat. Overall, the results presented in Tables 3.6 and 3.7 for cultivated area and Dataset 2 indicate that the UA and PA for the majority of the crop classes are near 1.

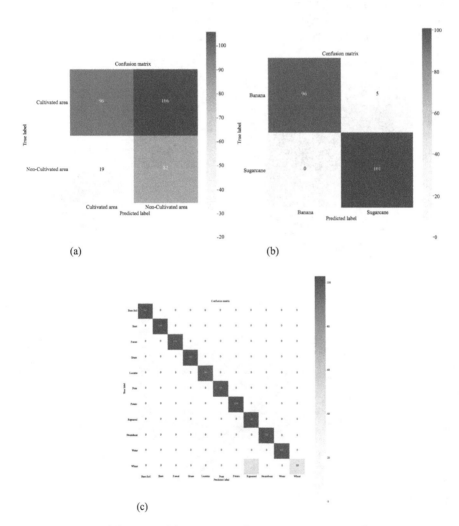

(a)

(b)

(c)

Figure 3.8   ROC curve: (a) Dataset 1, (b) cultivated area, and (c) Dataset 2

Table 3.6 Best user's (UA), producer's (PA), and overall (OA)
accuracy and kappa coefficient for cultivated area

Class	UA	PA	OA	Kappa
Banana	100%	95.04%	97.52%	0.95
Sugarcane	95.28%	100%		

Table 3.7 Best user's (UA), producer's (PA), and overall (OA)
accuracy and kappa coefficient for Dataset 2

Class	UA	PA	OA	Kappa
Bare soil	100%	100%		
Beet	100%	100%		
Forest	100%	100%		
Grass	98.07%	100%		
Lucerne	100%	98.01%		
Peas	100%	100%	95.17%	0.94
Potato	100%	100%		
Rapeseed	66.23%	100%		
Steambeans	100%	100%		
Water	100%	100%		
Wheat	100%	49.01%		

3.7 Conclusion

In this study, full polarization SAR data for crop classification in Bardoli and Flevoland was investigated. A deep learning-based crop classification method was used. Two types of approaches were used in this paper. One is Inception v3 and the other is Custom VGG like a model. Datasets were created for training a model followed by extracting fields of SAR images by super-imposing the edge detected optical/SAR image on the Freeman decomposition image and labeled extracted SAR field images for crop classification. The model achieved an overall accuracy of 97.52% for the cultivated area using Inception v3 and 95.17% for Dataset 2 using Custom VGG like model.

References

[1]   Atmanirbhar Krishi and Atmanirbhar Kisan are important to achieve the goal of Atmanirbhar Bharat –Prime Minister Shri Narendra Modi, in his address to the nation on the 74th Independence Day. Available from: https://pib.gov.in/PressReleasePage.aspx?PRID=1646286.

[2]   Ministry of Agriculture Farmers Welfare, Government of India. Kurukshetra Magazine: Ministry of Rural Development. Kurukshetra Magazine: Ministry

of Rural Development. Govt. of India 53, no. 1 (2004): 28. P. B. Patil and T. A. Shivare.

[3]   Take action for the sustainable development goals – United Nations sustainable development. United Nations. Available from: https://www.un.org/sustainabledevelopment/sustainable-development-goals.

[4]   Cué La Rosa LE, Queiroz Feitosa R, Nigri Happ P, *et al.* Combining deep learning and prior knowledge for crop mapping in tropical regions from multitemporal SAR image sequences. *Remote Sensing.* 2019;11(17):2029.

[5]   Sun Y, Luo J, Wu T, *et al.* Geo-parcel based crops classification with Sentinel-1 time series data via recurrent neural network. In: *2019 8th International Conference on Agro-Geoinformatics (Agro-Geoinformatics).* IEEE; 2019. p. 1–5.

[6]   Jia K, Li Q, Tian Y, *et al.* Crop classification using multi-configuration SAR data in the North China Plain. *International Journal of Remote Sensing.* 2012;33(1):170–183.

[7]   Kussul N, Skakun S, Shelestov A, *et al.* The use of satellite SAR imagery to crop classification in Ukraine within JECAM project. In: *2014 IEEE Geoscience and Remote Sensing Symposium.* IEEE; 2014. p. 1497–1500.

[8]   Tan CP, Ewe HT, and Chuah HT. Agricultural crop-type classification of multi-polarization SAR images using a hybrid entropy decomposition and support vector machine technique. *International Journal of Remote Sensing.* 2011;32(22):7057–7071.

[9]   Haldar D, Das A, Mohan S, *et al.* Assessment of L-band SAR data at different polarization combinations for crop and other land use classification. *Progress in Electromagnetics Research B.* 2012;36:303–321.

[10]   Skriver H. Crop classification by multitemporal C-and L-band single-and dual-polarization and fully polarimetric SAR. *IEEE Transactions on Geoscience and Remote Sensing.* 2011;50(6):2138–2149.

[11]   Li H, Zhang C, Zhang S, *et al.* Crop classification from full-year fully-polarimetric L-band UAVSAR time-series using the Random Forest algorithm. *International Journal of Applied Earth Observation and Geoinformation.* 2020;87:102032.

[12]   Denize J, Hubert-Moy L, and Pottier E. Polarimetric SAR time-series for identification of winter land use. *Sensors.* 2019;19(24):5574.

[13]   Zhao H, Chen Z, Jiang H, *et al.* Evaluation of three deep learning models for early crop classification using sentinel-1A imagery time series: a case study in Zhanjiang, China. *Remote Sensing.* 2019;11(22):2673.

[14]   Mullissa AG, Persello C, and Tolpekin V. Fully convolutional networks for multi-temporal SAR image classification. In: *IGARSS 2018–2018 IEEE International Geoscience and Remote Sensing Symposium.* IEEE; 2018. p. 6635–6638.

[15]   Xu Z, Meng S, Zhong S, *et al.* Study on temporal and spatial adaptability of crop classification models. In: *2019 8th International Conference on Agro-Geoinformatics (Agro-Geoinformatics).* IEEE; 2019. p. 1–7.

[16]   Mei X, Nie W, Liu J, *et al.* Polsar image crop classification based on deep residual learning network. In: *2018 7th International Conference on Agrogeoinformatics (Agro-geoinformatics)*. IEEE; 2018. p. 1–6.

[17]   Castro JDB, Feitoza RQ, La Rosa LC, *et al.* A comparative analysis of deep learning techniques for sub-tropical crop types recognition from multi-temporal optical/SAR image sequences. In: *2017 30th SIBGRAPI Conference on Graphics, Patterns and Images (SIBGRAPI)*. IEEE; 2017. p. 382–389.

[18]   Gu L, He F, and Yang S. Crop classification based on deep learning in northeast China using sar and optical imagery. In: 2019 *SAR in Big Data Era (BIGSARDATA)*. IEEE; 2019. p. 1–4.

[19]   Avolio C, Tricomi A, Mammone C, *et al.* A deep learning architecture for heterogeneous and irregularly sampled remote sensing time series. In: *IGARSS 2019 – 2019 IEEE International Geoscience and Remote Sensing Symposium*. IEEE; 2019. p. 9807–9810.

[20]   Kussul N, Lavreniuk M, Skakun S, *et al.* Deep learning classification of land cover and crop types using remote sensing data. *IEEE Geoscience and Remote Sensing Letters*. 2017;14(5):778–782.

[21]   Zhou Y, Luo J, Feng L, *et al.* Long-short-term-memory-based crop classification using high-resolution optical images and multi-temporal SAR data. *GIScience & Remote Sensing*. 2019;56(8):1170–1191.

[22]   Wei S, Zhang H, Wang C, *et al.* Multi-temporal SAR data large-scale crop mapping based on U-Net model. *Remote Sensing*. 2019;11(1):68.

[23]   Chen SW and Tao CS. PolSAR image classification using polarimetric-feature-driven deep convolutional neural network. *IEEE Geoscience and Remote Sensing Letters*. 2018;15(4):627–631.

[24]   Chen SW and Tao CS. Multi-temporal PolSAR crops classification using polarimetric-feature-driven deep convolutional neural network. In: *2017 International Workshop on Remote Sensing with Intelligent Processing (RSIP)*. IEEE; 2017. p. 1–4.

[25]   Ndikumana E, Ho Tong Minh D, Baghdadi N, *et al.* Deep recurrent neural network for agricultural classification using multitemporal SAR Sentinel-1 for Camargue, France. *Remote Sensing*. 2018;10(8):1217.

[26]   Xing Y, Zhang Y, Li N, *et al.* Improved superpixel-based polarimetric synthetic aperture radar image classification integrating color features. *Journal of Applied Remote Sensing*. 2016;10(2):026026.

[27]   Dong H, Zou B, Zhang L, *et al.* Automatic design of CNNs via differentiable neural architecture search for PolSAR image classification. *IEEE Transactions on Geoscience and Remote Sensing*. 2020;58(9):6362–6375.

[28]   He C, He B, Tu M, *et al.* Fully convolutional networks and a manifold graph embedding-based algorithm for polsar image classification. *Remote Sensing*. 2020;12(9):1467.

[29]   European Space Agency. Available from: https://earth.esa.int/documents/653194/656796/LNAdvancedConcepts.pdf.

[30]   Xie S and Tu Z. Holistically-nested edge detection. In: *Proceedings of the IEEE International Conference on Computer Vision*; 2015. p. 1395–1403.

[31]   Python Tutorials. Available from: https://opencv-python-tutroals.read-thedocs.io/en/latest/pytutorials/pyimgproc/pycanny/pycanny.html.

[32]   Python Filtering. Available from: https://opencv-python-tutroals.readthedocs.io/en/latest/pytutorials/pyimgproc/pyfiltering/pyfiltering.html.

[33]   Russakovsky O, Deng J, Su H, *et al.* Imagenet large scale visual recognition challenge. *International Journal of Computer Vision.* 2015;115(3):211–252.

[34]   Tang Y, Li X, Jiang Y, *et al.* Research on human eye iris recognition based on inception v3. *Frontiers in Signal Processing.* 2019;3(4):34001.

[35]   Sun X and Wei J. Identification of maize disease based on transfer learning. *Journal of Physics*: *Conference Series.* 2020;1437:012080.

[36]   Bai J, Wang B, Chen C, *et al.* Inception-v3 based method of LifeCLEF2019 bird recognition. In: *CLEF (Working Notes)*; 2019.

[37]   Kundu S and Maulik U. Vehicle pollution detection from images using deep learning. In: *Intelligence Enabled Research.* New York, NY: Springer; 2020. p. 1–5.

[38]   Saini M and Susan S. Data augmentation of minority class with transfer learning for classification of imbalanced breast cancer dataset using inception-v3. In: *Iberian Conference on Pattern Recognition and Image Analysis.* New York, NY: Springer; 2019. p. 409–420.

[39]   Varaich ZA and Khalid S. Recognizing actions of distracted drivers using inception v3 and exception convolutional neural networks. In: *2019 2nd International Conference on Advancements in Computational Sciences (ICACS).* IEEE; 2019. p. 1–8.

[40]   Santhiya DS, Pravallika S, Sukrutha MA, *et al.* Skin disease detection using v2 and v3 in machine learning. *International Journal of Engineering Science.* 2019;9:21343.

[41]   Xia X, Xu C, and Nan B. Inception-v3 for flower classification. In: *2017 2nd International Conference on Image, Vision and Computing (ICIVC).* IEEE; 2017. p. 783–787.

[42]   Pires de Lima R and Marfurt K. Convolutional neural network for remote-sensing scene classification: transfer learning analysis. *Remote Sensing.* 2019;12(1):86.

[43]   Szegedy C, Vanhoucke V, Ioffe S, *et al.* Rethinking the inception architecture for computer vision. In: *Proceedings of the IEEE Conference on Computer Vision and Pattern Recognition*; 2016. p. 2818–2826.

[44]   Simonyan K and Zisserman A. Very deep convolutional networks for large-scale image recognition; 2014. arXiv preprint arXiv:14091556.

[45]   Bardoli. Wikimedia Foundation; 2022. Available from: https://en.wikipedia.org/wiki/Bardoli.

[46]   Step. Available from: https://earth.esa.int/web/polsarpro/data-sources/sample-datasets.

Chapter 4

Leveraging twin networks for land use land cover classification

*Pranshav Gajjar[1], Manav Garg[1], Pooja Shah[2],
Vijay Ukani[1] and Anup Das[3]*

Information related to land use land cover (LULC) plays an instrumental role in land management and planning. With the advent of the field of machine learning, accurate automation of tasks has become feasible, hence this study presents a similarity-learning approach using twin networks for LULC classification and extended use cases on the AVIRIS sensor's Indian Pines standard dataset. A thorough comparative study is conducted for the Siamese Network backbones, with experiments on DiCENets, ResNets, SqueezeNets, and related state-of-the-art approaches. Embedding augmentation is also explored along with embedding classification and dimensionality reduction algorithms for understanding hyperspace generation and the use of similarity learning. The performance analysis of SiameseNets on a reduced training dataset size is also shown to reinforce the utility of SiameseNets. Thorough experiments are conducted for improving the hyperparameters associated with deep learning architectures to form a non-biased and rational comparison of the given classification approaches. The proposed methodologies follow a classification accuracy approaching 98% and are validated using baselines.

4.1 Introduction

Accurately and efficiently identifying the physical characteristics of the earth's surface or land cover in general, as well as how we exploit the affiliated land usage, is a difficult topic in environmental monitoring and many other related areas and subdomains [1]. This can be accomplished by extensive field surveys or by analyzing a large number of satellite photos utilizing the remote sensing paradigm [2]. While doing field surveys is more complete and authoritative, it is a costly

[1]Institute of Technology, Nirma University, India
[2]School of Technology, Pandit Deendayal Energy University, India
[3]Space Application Centre, ISRO, India

endeavor that often takes a long time to update. Deep learning and convolutional neural networks have demonstrated promising results in LULC categorization with recent breakthroughs in the space industry and the growing availability of satellite pictures in both free and commercially accessible data sets [3]. By assessing and understanding the current status of predictive analysis and general research related to vision systems, application-oriented deep learning [4–6], computing, computational facilities [7], and the overall implications of robust remote sensing, this chapter offers novel implementations and analysis in the field of machine learning-assisted LULC predictions. For any predictive model or any algorithm related to the machine learning methodology, an inherent need for sufficient data is encountered [8]. It is the general thought process and observation around deep learning algorithms that a higher number of data samples implicates a higher possible classification or prediction accuracy [8].

However, in cases concerning a multitude of domains a deficiency of significant data samples is observed, and special contingencies should be anointed to tackle related predicaments. Recent literature and advancements in the field of deep learning recommend two possibilities to tackle an insufficient dataset, namely, data augmentation [9], and similarity learning [10] implementations. To elaborate details on the former, there have been significant improvements in the field of data augmentations, with algorithms focusing on images, Generative Adversarial Networks (GANs) [11], or the excessively contemporary embedding-based augmentative strategies [12]. There have been many image augmentation techniques in the literature on images, they usually promote the use of image array manipulations like randomized rotations, horizontal flipping, and vertical flipping [8]. However, the computation corresponding to image manipulation would be significantly higher when compared to embeddings or moderately dimensionalized embedding data. The latter argumentation mechanism for scarce data is countered by the use of Siamese Networks or similarity learning, these specialized deep neural architectures permit the use of learning similar features from pairs, triplets, or quadruplets and traverse the information to favorable embeddings and embedded hyperspace [13]. With the scope of this paper related to the use of Siamese networks, an E-Mixup embedding augmentation strategy [12] is presented to accurately measure and validate the use of Siamese networks, augmentation, and its associated prominence for LULC classification. This chapter functions on the motivations related to automating the task of LULC classification and the data-related predicaments, and functions on a publicly available Indian Pines dataset, which consists of AVIRIS Hyperspectral images [14]. The chapter further contains the related work, the proposed computation strategies and methodologies, and a thorough empirical analysis to understand the relevance of the mentioned literature followed by the concluding statements.

4.2   Related literature

The current literature offers a significant understanding of computation and how novel ideations concerning machine learning are operated and connected to remote sensing and LULC classification and general predictive analysis of LULC. A novel analysis presented in [15] showcases a deeper insight into the domain of medical imaging and the relevant machine learning and deep learning algorithms which assisted the related cause. The paper offered a similar correlation between the use of deep learning algorithms, and the challenges, possibilities, and benchmarks corresponding to remote sensing. The article [16] offered a novel analysis of the spatial relationships between land surface temperature, commonly abbreviated as LST, and LULC of three urban agglomerations (UAs) belonging to the eastern portion of India. The obtained study further promoted a planning and implementation strategy for the effective development of small and medium-sized cities. The paper [17] presented a comparative study, where different object-based and pixel-based classification technologies were leveraged on a high spatial-resolution multi-source dataset to accurately and robustly map LULC. Significant work has also been done for learning transferable change for LULC change detection, the novel method as presented in [18] had significant societal information and added to the multi-disciplinary utility of an RNN. The authors of the paper [14] presented and highlighted the use and functioning of a convolutional neural network (CNN) in a study area focusing on Aurangabad, India. The presented model procured a percentage accuracy of 79.43 when validated against the previously mentioned study area. An important aspect concerning the algorithm was the use of CNNs for an unstructured input and a significantly scarce data pool, as the model performance turned out to be satisfactory. The previously mentioned research also utilized the Indian Pines dataset, further explaining and justifying its usefulness for this chapter. The paper [19] also functioned primarily on deep neural networks, however, offered a different study area and learning paradigm. The related deep neural architectures were trained using the publicly available ImageNet LargeScale Visual Recognition Competition (ILSVRC) datasets. The models worked on the fine-tuning principle and a total of 19,000 Landsat 5/7 satellite images from the Province of Manitoba in Canada were converted to patches of image size 224 × 224 for the same commission. The article [20] also worked on LULC classification on data collected by the Sentinel-2, the dataset collection is called EuroSAT remote sensing image dataset and was used for the underlying tests. Several CNN-based subsidiary architectures like InceptionV3, ResNet50, and VGG19 were deployed for feature extraction and methodologies containing Channel Squeeze & Spatial Excitation block and Twin SVM (TWSVM) were further used as a complete pipeline. The best performing model outputted a 94.57% accuracy.

4.3 Methodology

The chapter offers a novel study on deep architectures, Siamese or Twin networks, and classification approaches to the Indian Pines dataset. Multiple experiments were conducted, where the encoder was tested without any modifications, a Siamese variant for the encoder, and the siamese variant with E-Mixup augmentation. This section additionally contains the dataset description, information about

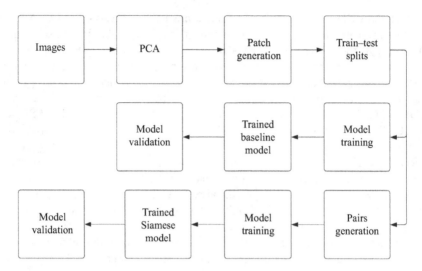

Figure 4.1   Information flow and the primary preprocessing steps for training and testing the proposed models, each architecture, and underlying technologies are explained thoroughly in the forthcoming sections.

Siamese Networks and the related parameters, and a thorough description of the experimented encoders. The primary flow of information which is observed for training and validating the proposed neural methodologies is explained in Figure 4.1.

4.3.1   Dataset

For this study, the Indian Pines dataset was used, and the underlying preprocessing steps are mentioned and explained in this subsection. The images follow 145 × 145 spatial dimensions and 224 spectral bands in the wavelength range of 400–2,500 nm. A total of 24 spectral bands concerning water absorption regions have been discarded. The available ground truth images are designated to 16 different classes which adequately categorizes the study area's land cover. The images have a significantly high input dimensionality which is unsuited for training, hence Principal Component Analysis or PCA [21] is deployed. The resultant is 10,249 images of size 64 × 64 × 3, the totality of the time taken to perform the said conversion was 0.616 s. To avoid the skewed dataset problem, a stratified train test split [22] is used, and an equal percentage of images are taken out of each class to have a high-quality training setup. The standard encoders or single architectures are trained on a 60–40 train-test split and to check the SiameseNet's performance on a reduced training test size a 50–50 split is leveraged. The lower train–test split was added to future-proof LULC studies, as for cases about the inclusion of a new class, data sample-based predicaments may be observed, and having standardized and developed neural strategies which function suitably in those scenarios, is highly favorable. A tabular description for the 16 constituent classes [23] is mentioned in Table 4.1.

Table 4.1   The frequency distribution for the
16 LULC classes [23]

Sr. no.	Class	Samples
1	Soybean-clean	593
2	Alfafa	46
3	Corn-mintill	830
4	Grass-pasture-mowed	28
5	Grass-pasture	483
6	Grass-trees	730
7	Hay-windrowed	478
8	Corn	237
9	Stone-steel-towers	93
10	Woods	972
11	Soybean-mintill	2,455
12	Corn-notill	1,428
13	Wheat	205
14	Soybean-notill	972
15	Building-grass-trees-drives	386
16	Oats	20

4.3.2   Siamese network

For predicting the similarity index between input pairs, a Siamese network with twin convolutional neural networks is trained. Because the network parameters are the same, the model learns the same transformations, resulting in a favorable feature space. The base networks are trained with the Binary Cross-Entropy loss function [24], which is generated by utilizing the representations learned at the previous fully connected layer. The basic network is utilized to generate embeddings, which are then categorized by an multi-layer perceptron (MLP) [25] and enhanced and augmented using the E-Mixup methodology.

4.3.2.1   E-Mixup

This section elaborates on the embedding augmentation functionality as proposed in the aforementioned sections. The embeddings that are generated due to the use of a Siamese Network have a significantly lower size than a standard image, hence using an embedding augmentation technique generally outperforms standard image manipulation techniques in terms of computational efficiency. To use the E-Mixup augmentation pipeline, a weighted average is taken over two embeddings, and the binary matrix representation of the class values with the weight as lambda [12], which is calculated as a random value from a Beta distribution with the alpha values fixed as 0.2 [12]. By using the said method, the training set is tripled and classified.

4.3.2.2   MLP

This section elaborates on the further classification of the obtained embeddings as proposed in the aforementioned sections. The embeddings have a dimensionality of

256 and are fed into an MLP. The architecture contains four hidden layers, batch normalization, and the concluding Softmax layer. The final activation function promotes a resultant probabilistic distribution, which is favorable for a multiclass classification problem. The same architecture is used for all experiments related to Siamese Networks, for an unbiased and nonaligned testing experience.

4.3.3   Encoders

Various encoders architectures were tested, mainly DenseNets [26], DiceNets [27], ResNets [28], and SqueezeNets [29]. For tests involving individual networks, a softmax layer is used for the classification. For tests about Siamese Networks, each encoder is modified to obtain embeddings with 256 dimensions by using dense layers.

4.3.3.1   DiCENet

Standard convolutions simultaneously encode spatial and channel-wise information, but they are computationally intensive. Therefore, separable (or depth-wise separable) convolutions are introduced to increase the efficiency of ordinary convolutions by encoding spatial and channel-wise information independently using depth-wise and point-wise convolutions, respectively. Though successful, this factorization places a large computational burden on point-wise convolutions, making them a computational bottleneck. The DiCENet consists of multiple DiCE units, which can be understood as building blocks [30].

The DiCE unit was introduced to encode spatial and channel-wise information efficiently. Dimensionwise Convolution (DimConv) and Dimension-wise Fusion are used to factorize ordinary convolution in the DiCE unit (DimFuse). DimConv learns local dimension-wise representations by applying light filtering across each dimension of the input tensor. DimFuse effectively mixes these representations from many sizes while also incorporating global data. The DiCE unit can employ DimFuse instead of computationally intensive point-wise convolutions since DimConv can encode local spatial and channel-wise information from all dimensions.

4.3.3.2   ResNet

This item presents a synopsis of ResNets as illustrated in the original publication [28]. One of the proposed encoders that we will utilize is Residual Networks, often known as ResNets. The advantage that ResNets have over unadorned networks is that they alleviate the deterioration issue that was shown when extremely deep networks began to converge. Identity mapping was presented by ResNets, which meant that input from a prior layer was taken and sent to another layer as a shortcut. A deeper understanding can be obtained from Figure 4.2 [28].

Typically, the 34-layer and 18-layer ResNets are employed because they produce a substantially small error and comparably higher accuracies than their simple rivals. The 34-layer ResNet exhibits somewhat lower training error and tackles the deterioration problem encountered in its plain competitor, resulting in good accuracy from greater depths. Not only did the 18-layer ResNet outperform its simple competitor in terms of accuracy, but it was also able to perpetrate convergence faster and get satisfactory solutions on fewer data entities.

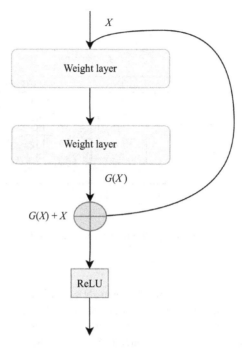

Figure 4.2   Underlying functionality of the residual architecture [28]

4.3.3.3   SqueezeNet

SqueezeNet's research involves an intelligent architecture as well as a quantitative analysis. As a consequence, while maintaining the same degree of accuracy, SqueezeNet can be three times quicker and 500 times smaller than AlexNet. The fire module is a two-layer SqueezeNet building block featuring a squeeze layer and an expansion layer. A SqueezeNet is a network made up of many fire modules and numerous pooling layers. The squeeze and expand layers keep the feature map size constant, while the former decreases the depth while the latter increases it [29]. Another common design is to increase the depth while lowering the breadth. Because the squeeze module includes just 1×1 filters, it acts as a fully-connected layer that operates on feature points in the exact position. In other words, it cannot be spatially abstracted. As the name says, one of its advantages is that it minimizes the depth of the feature map. When the depth is lowered, the succeeding 3×3 filters in the expansion layer will have more undersized calculations [29]. It boosts performance since a 3×3 filter takes nine times the calculation of a 1×1 filter. Intuitively, too much squeezing inhibits information flow; too few 3×3 filters, on the other hand, restrict spatial resolution. These were the primary architectural details that were obtained from the original article [29]. The inherent functioning of this architecture and fire modules is explained graphically in Figure 4.3 [31].

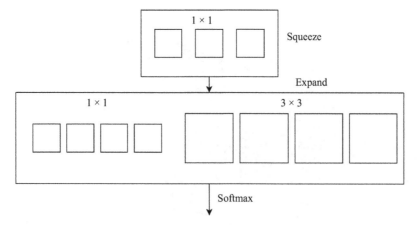

Figure 4.3   Functionality of a fire module, as depicted in the research [31]

4.3.3.4   DenseNet

This section presents a synopsis of the DenseNet architecture as specified in the original paper [26]. Another type of encoder employed in the suggested design is the Dense Convolution Network, often known as DenseNet. The benefit of employing this sort of network is that each of its levels collects supplemental inputs from all of the layers before it. The data is concatenated such that each layer acquires the cumulative intelligence of all preceding levels. As a result, when each layer acquires feature mappings from previous layers, the whole network is compressed, resulting in more infrequent total channels. The constituent functionality of the DenseNet neural architecture can be further understood by Figure 4.4 [26].

The difference between DenseNets and ResNets is that DenseNets employ the parameters more dexterously. Outwardly, the only significant difference between the two networks is that DenseNets concatenate the inputs, whereas ResNets sum the inputs. Although this appears to be a little modification, it results in a significant shift in behavior for both of them [26]. Furthermore, DenseNets require a small number of parameters and processing resources to deliver extremely precise and cutting-edge findings, and results with greater precision may be obtained when the hyperparameters are tweaked with care [26].

4.4   Results and discussion

This section contains the empirical data associated with the conducted experiments. Each model was trained for sufficient epochs and the best performing entity was obtained through extensive callbacks. The various deep neural architectures were deployed primarily using Keras [32] and PyTorch [33]. The performance of each tested strategy was assessed by using a plethora of performance metrics, the chapter leverages the F1 score, precision, recall, percent accuracy [34], and Cohen's kappa coefficient [35]. The primary metrics can be further explained analytically as the

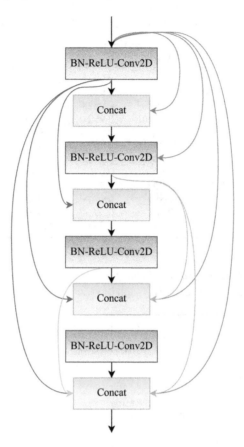

Figure 4.4   Underlying functionality of the DenseNet neural architecture [26]

following equations, here TP, FP, TN, FN implicates the True Positive, False Positive, True Negative, and True Positive respectively:

$$F1\text{-}Score = \frac{2 * TP}{2 * TP + FP + FN} \tag{4.1}$$

$$Accuracy = \frac{TP + TN}{TP + TN + FP + FN} \tag{4.2}$$

The obtained results concerning the aforementioned performance metrics are mentioned in Table 4.2. A substantial temporal analysis is also obtained by assessing the time taken for both the training phases and the testing phases across all experiments for understanding the deploy-ability of each considered methodology is mentioned in Table 4.3.

From Tables 4.2 and 4.3, it can be inferred that the best performing models were obtained during the tests concerning Siamese DenseNets, which were augmented using E-Mixup and the standard ResNet. To further estimate the best

Table 4.2   Results concerning the primary performance metrics

Model	Accuracy	F1-score	Kappa	Precision	Recall
ResNet	99.707	0.997	0.996	0.997	0.997
ResNet-S + MLP	96.19	0.962	0.956	0.963	0.962
ResNet-S + E-M + MLP	96.26	0.9624	0.957	0.964	0.963
DiCENet	96.17	0.962	0.956	0.963	0.962
DiCENet-S + MLP	91.79	0.917	0.906	0.918	0.917
DiCENet-S + E-M + MLP	90.85	0.907	0.896	0.907	0.908
DenseNet	99.56	0.996	0.995	0.996	0.996
DenseNet-S + MLP	99.66	0.997	0.996	0.997	0.996
DenseNet-S + E-M + MLP	99.707	0.997	0.997	0.996	0.997
SqueezeNet	97.881	0.971	0.976	0.963	0.978
SqueezeNet-S + MLP	99.001	0.99	0.989	0.991	0.99
SqueezeNet-S + E-M + MLP	99.34	0.993	0.992	0.993	0.993

Table 4.3   Results associated with the temporal characteristics, each unit is in seconds

Model	Training time (s)	Prediction time (s)
ResNet	86.864	0.894
ResNet-S + MLP	315.681	2.638
ResNet-S + E-M + MLP	316,278	2.638
DiCENet	54.777	1.353
DiCENet-S + MLP	84.199	2.659
DiCENet-S + E-M + MLP	94.895	2.659
DenseNet	113.321	1.31
DenseNet-S + MLP	275.306	0.6912
DenseNet-S + E-M + MLP	276.438	0.6912
SqueezeNet	14.111	0.333
SqueezeNet-S + MLP	70.946	0.732
SqueezeNet-S + E-M + MLP	107.219	0.732

performing model, a Standard ResNet supersedes the DenseNet configuration due to its computationally efficient inference. The DenseNet also has its uses for experimental setups permitting a longer training duration, and implicating a sparse dataset, using this architectural combination guarantees an identical performance. The best performing model, when all the performance metrics are leveraged and a trade-off is obtained with computational efficiency, is the Siamese SqueezeNet with E-Mixup. The utility of Siamese enhancements can also be seen for the SqueezeNet, as the twin networks outperformed the standard variant. To further understand the predictions of these models, the below-mentioned figure offers a graphical comparison between all the 12 architectural test cases. The limitations in the predictive performances for some architectures can be observed.

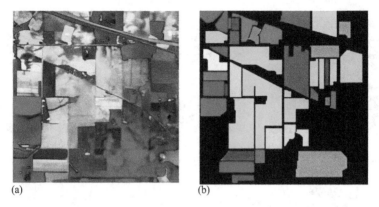

(a)   (b)

*Figure 4.5   The false-color composite (a) and the corresponding ground truth
image (b)*

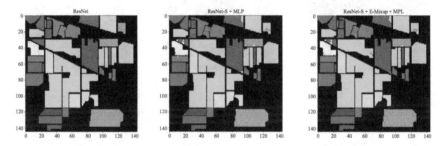

*Figure 4.6   The obtained spectral graphs for the ResNet-based architectures, the
architectural names are based on the aforementioned abbreviations.*

4.5   Conclusion and future work

This chapter aimed to offer a novel implementation and study for possible advancements
in the field of LULC classification. The Indian Pines dataset was used for training and
validation of the mentioned methodologies, with extensive tests on DiceNets, DenseNets,
ResNets, and SqueezeNets. The individual encoders were also tested in a low train test
split with a twin network strategy, which symbolized and paved the way for further
research involving a potential new class, where data scarcity is common. An embedding
augmentation strategy called E-Mixup or Embedding-Mixup is also explored which was
emphasized as an improvement to the standard or vanilla Siamese implementations.

For future work, the authors aim to improve upon the existing datasets and also
work on LULC predictions about Indian terrains to validate the feasibility and
traversability of the Indian Pines Dataset which can be considered a limitation for
the scope of this study. The authors also aim to assess other modern architectural
strategies and newer learning techniques to present a potentially generalized study.

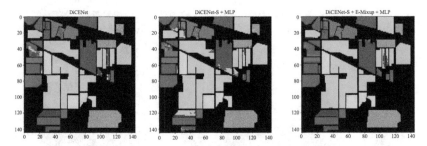

*Figure 4.7   The obtained spectral graphs for the DiCENet-based architectures,
the architectural names are based on the aforementioned
abbreviations.*

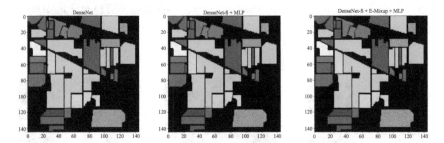

*Figure 4.8   The obtained spectral graphs for the DenseNet-based architectures,
the architectural names are based on the aforementioned
abbreviations.*

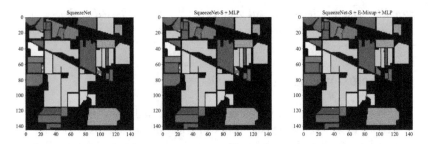

*Figure 4.9   The obtained spectral graphs for the SqueezeNet-based architectures,
the architectural names are based on the aforementioned
abbreviations.*

References

[1]   Sathyanarayanan D, Anudeep D, Das CAK, *et al*. A multiclass deep learning
approach for LULC classification of multispectral satellite images. In: *2020
IEEE India Geoscience and Remote Sensing Symposium (InGARSS)*. IEEE;
2020. Available from: http://dx.doi.org/10.1109/ingarss48198.2020.9358947.

[2] Naushad R, Kaur T, and Ghaderpour E. Deep transfer learning for land use and land cover classification: a comparative study. *Sensors* 2021;21 (23):8083. Available from: http://arxiv.org/abs/2110.02580v3.

[3] Yassine H, Tout K, and Jaber M. Improving LULC classification from satellite imagery using deep learning – EUROSAT Dataset. In: *Conference: The XXIV ISPRS Congress (TCIII-Remote Sensing)*, 2021, vol. 6, p. XLIII-B3-2021:369–376. Available from: http://dx.doi.org/10.5194/isprs-archives-xliii-b3-2021-369-2021.

[4] Gajjar P, Shah P, and Sanghvi H. E-Mixup and Siamese Networks for musical key estimation. In: *Smart Innovation, Systems and Technologies*. Singapore: Springer Nature; 2022. p. 343–350. Available from: http://dx.doi.org/10.1007/978-981-19-2541-2_26.

[5] Gajjar P, Garg M, Shah V, *et al.* Applicability analysis of attention U-Nets over vanilla variants for automated ship detection. *Reports on Geodesy and Geoinformatics*. 2022;114(1):9–14.

[6] Gajjar P, Mehta N, and Shah P. Quadruplet loss and SqueezeNets for Covid-19 detection from chest-X ray. *CSJM* 2022;30(2(89)):214–222. Available from: http://dx.doi.org/10.56415/csjm.v30.12.

[7] Mehta N, Shah P, Gajjar P, *et al.* Ocean surface pollution detection: applicability analysis of V-Net with data augmentation for oil spill and other related ocean surface feature monitoring. In: *Communication and Intelligent Systems*. Singapore: Springer Nature; 2022. p. 11–25. Available from: http://dx.doi.org/10.1007/978-981-19-2130-8_2.

[8] Mehta N, Shah P, and Gajjar P. Oil spill detection over ocean surface using deep learning: a comparative study. *Marine Systems & Ocean Technology*, 2021 11;16(3–4):213–220. Available from: http://dx.doi.org/10.1007/s40868-021-00109-4.

[9] He Z, Xie L, Chen X, Zhang Y, Wang Y, and Tian Q. Data augmentation revisited: rethinking the distribution gap between clean and augmented data. arXiv preprint arXiv:1909.09148. 2019. Available from: http://arxiv.org/abs/1909.09148v2.

[10] Qiu K, Ai Y, Tian B, *et al.* Siamese-ResNet: implementing loop closure detection based on Siamese network. In: *2018 IEEE Intelligent Vehicles Symposium (IV)*. IEEE; 2018. Available from: http://dx.doi.org/10.1109/ivs.2018.8500465.

[11] Ghojogh B, Ghodsi A, Karray F, and Crowley M. Generative Adversarial Networks and Adversarial Autoencoders: Tutorial and Survey. arXiv preprint arXiv:2111.13282. 2021. Available from: http://arxiv.org/abs/2111.13282v1.

[12] Wolfe CR and Lundgaard KT. E-Stitchup: Data Augmentation for Pre-Trained Embeddings. arXiv preprint arXiv:1912.00772. 2019. Available from: http://arxiv.org/abs/1912.00772v2.

[13] Shrestha A and Mahmood A. Enhancing Siamese networks training with importance sampling. In: *Proceedings of the 11th International Conference on Agents and Artificial Intelligence*. SCITEPRESS – Science and Technology

Publications; 2019. Available from: http://dx.doi.org/10.5220/000737170 6100615.

[14] Bhosle K and Musande V. Evaluation of deep learning CNN model for land use land cover classification and crop identification using hyperspectral remote sensing images. *Journal of the Indian Society of Remote Sensing*, 2019;47(11),1949–1958. Available from: http://dx.doi.org/10.1007/s12524-019-01041-2.

[15] Viswanathan J, Saranya N, and Inbamani A. Deep learning applications in medical imaging. In: *Deep Learning Applications in Medical Imaging*. IGI Global; 2021. p. 156–177. Available from: http://dx.doi.org/10.4018/978-1-7998-5071-7.ch007.

[16] Saha S, Saha A, Das M, Saha A, Sarkar R, and Das A. Analyzing spatial relationship between land use/land cover (LULC) and land surface temperature (LST) of three urban agglomerations (UAs) of Eastern India. *Remote Sensing Applications: Society and Environment*, 2021;22,100507. Available from: http://dx.doi.org/10.1016/j.rsase.2021.100507.

[17] Balha A, Mallick J, Pandey S, Gupta S, and Singh CK. A comparative analysis of different pixel and object-based classification algorithms using multi-source high spatial resolution satellite data for LULC mapping. *Earth Science Informatics*, 2021;14(4),2231–2247. Available from: http://dx.doi.org/10.1007/s12145-021-00685-4.

[18] Lyu H, Lu H, and Mou L. Learning a transferable change rule from a recurrent neural network for land cover change detection. *Remote Sensing*, 2016;8(6),506. Available from: http://dx.doi.org/10.3390/rs8060506.

[19] Henry CJ, Storie CD, Palaniappan M, *et al.* Automated LULC map production using deep neural networks. *International Journal of Remote Sensing*, 2019;40(11),4416–4440. Available from: http://dx.doi.org/10.1080/01431161.2018.1563840.

[20] Dewangkoro HI and Arymurthy AM. Land use and land cover classification using CNN, SVM, and Channel squeeze & spatial excitation block. *IOP Conference Series: Earth and Environmental Science*, IOP Publishing 2021;704 (1):012048. Available from: http://dx.doi.org/10.1088/1755-1315/704/1/012048.

[21] Jiang H and Meyer C. Modularity component analysis versus principal component analysis. arXiv preprint arXiv:1510.05492. 2015. Available from: http://arxiv.org/abs/1510.05492v2.

[22] Farias F, Ludermir T, and Bastos-Filho C. Similarity based stratified splitting: an approach to train better classifiers. 2020. Available from: http://arxiv.org/abs/2010.06099v1.

[23] Gao Q, Lim S, and Jia X. Hyperspectral image classification using convolutional neural networks and multiple feature learning. *Remote Sensing*, 2018;10(2):299. Available from: http://dx.doi.org/10.3390/rs10020299.

[24] Andreieva V and Shvai N. Generalization of cross-entropy loss function for image classification. *Journal of Mathematical Sciences and Applications*, 2021; 1(3):3–10. doi: 10.18523/2617-7080320203-10, Available from: http://dx.doi.org/10.18523/2617-7080320203-10.

[25]   Shepanski JF. Multilayer perceptron training using optimal estimation. *Neural Networks*, 1988;1:51. Available from: http://dx.doi.org/10.1016/0893-6080 (88)90093-7.

[26]   Que Y and Lee HJ. Densely connected convolutional networks for multi-exposure fusion. In: 2018 *International Conference on Computational Science and Computational Intelligence (CSCI)*. IEEE; 2018. Available from: http://dx.doi.org/10.1109/csci46756.2018.00084.

[27]   Pal A, Krishnan G, Moorthy MR, *et al*. DICENet: fine-grained recognition via dilated iterative contextual encoding. In: *2019 International Joint Conference on Neural Networks (IJCNN)*. IEEE; 2019. Available from: http://dx.doi.org/10.1109/ijcnn.2019.8851800.

[28]   He K, Zhang X, Ren S, *et al*. Deep residual learning for image recognition. In: *2016 IEEE Conference on Computer Vision and Pattern Recognition (CVPR)*. IEEE; 2016. Available from: http://dx.doi.org/10.1109/cvpr.2016.90.

[29]   Iandola FN, Han S, Moskewicz MW, Ashraf K, Dally WJ, and Keutzer K. SqueezeNet: AlexNet-level accuracy with 50x fewer parameters and <0.5MB model size. 2016. arXiv preprint arXiv:1602.07360. Available from: http://arxiv.org/abs/1602.07360v4.

[30]   Metha S, Hajishirzi H, and Rastegari M. DiCENet: dimension-wise convolutions for efficient networks. *IEEE Transactions on Pattern Analysis and Machine Intelligence* 2020;44:2416–2425. Available from: https://dx.doi. org/10.1109/TPAMI.2020.3041871.

[31]   Wang A, Wang M, Jiang K, Cao M, and Iwahori Y. A dual neural architecture combined SqueezeNet with OctConv for LiDAR data classification. *Sensors*, 2019;19(22):4927. Available from: http://dx.doi.org/10.3390/ s19224927.

[32]   Reiser P, Eberhard A, and Friederich P. Implementing graph neural networks with TensorFlow-Keras. arXiv preprint arXiv:2103.04318. 2021. Available from: http://arxiv.org/abs/2103.04318v1.

[33]   Mishra P. Introduction to neural networks using PyTorch. In: *PyTorch Recipes. Apress*; 2019. p. 111–126. Available from: http://dx.doi.org/ 10.1007/978-1-4842-4258-2_4.

[34]   Bittrich S, Kaden M, Leberecht C, *et al*. Application of an interpretable classification model on Early Folding Residues during protein folding. *BioData mining*, 2019 1;12 1–16. Available from: http://dx.doi.org/10.1186/ s13040-018-0188-2.

[35]   Roy SK, Krishna G, Dubey SR, *et al*. HybridSN: exploring 3-D-2-D CNN feature hierarchy for hyperspectral image classification. *IEEE Geoscience and Remote Sensing Letters*, 2020;17(2):277–281. Available from: http://dx. doi.org/10.1109/lgrs.2019.2918719.

Chapter 5

Exploiting artificial immune networks for enhancing RS image classification

*Poonam S. Tiwari[1], Hina Pande[1] and
Shrushti S. Jadawala[2]*

The use of up-to-date and detailed information about the urban land cover is strategic for urban planning and management in present times, involving issues related to the recent massive urban sprawl and densification, climate change, and the need for environmental protection. The remote sensing images obtained by high spatial resolution satellite sensors are important data sources for urban classification. The artificial immune network (AIN), a computational intelligence model based on artificial immune systems (AIS) inspired by the vertebrate immune system, has been widely utilized for pattern recognition and data analysis. The algorithm is based on the principles of the behaviours of both B cells and T cells in the biological immune system. However, due to the inherent complexity of current AIN models, their application to multi-/hyperspectral remote sensing image classification has been severely restricted. The study explores the accuracy gained in land cover classification using the AIN.

The algorithm is inspired by the clonal selection theory of acquired immunity that explains how B and T lymphocytes improve their response to antigens over time. The region of interest (ROI) is selected. Clonal selection algorithms are most commonly applied to optimization and pattern recognition domains. The algorithm is initialised by randomly chosen pixels to a set of memory cells Ab{m} and to the set of Ab{r}. Euclidian distance is calculated to define the affinity to Ab. The cells showing the highest affinity are labelled as memory cells. The antibody cells are cloned and mutated, and finally the affinity is calculated, and memory cells are updated. The final classification is carried out based on affinity to the memory cells.

High-resolution earth observation data for Chandigarh city, India, has been utilised for the study. The urban land cover was extracted using the AIN algorithm and maximum likelihood classifier and a comparative analysis was carried out.

[1]Department of Space, Indian Institute of Remote Sensing, Indian Space Research Organization, Government of India, India
[2]Department of Earth Sciences, School of Science, Gujarat University, India

Data was classified into 15 urban landcover classes such as roads, roof types and vegetation. It was observed that for most of the classes, an improvement in overall accuracy and kappa statistics was observed by using the approach based on an AIN. Overall kappa statistics for MLC was found to be 0.72, and for AIN, it was calculated to be 0.86. The study demonstrates the potential of AIN-based algorithms for the classification of high-resolution remote sensing images.

5.1   Introduction

Due to an immense population, rapid urbanization and industrialization, land resources worldwide have been facing unprecedented pressure in the past decades, highlighted by the rapid loss of quality agricultural land and excessive sprawl of urban boundaries. The situation in countries with large populations, e.g. India is quite alarming [1–3]. Therefore there is an urgent need to create a balance between the land-use supply and demand in a region and promote the sustainable utilisation of land resources. This will lead to better land-use planning and optimisation of resources [1,4].

Image classification is an important issue in remote sensing and other applications. The accurate classification of remote sensing images has a wide range of uses, including reconnaissance, assessment of environmental damage, land use monitoring, urban planning and growth regulation [5–7]. A significant distinction in image classification separates supervised from unsupervised classification methods. In remote sensing image classification, a key issue is to improve classification accuracy. For many years, a conventional statistical classifier, such as maximum-likelihood (ML), has been applied for remote sensing image classification. However, these conventional multivariate statistical methods require non-singular and class-specific covariance matrices for all classes. Because of the complexity of ground matters and the diversity of disturbance, these traditional classification methods often have the drawback of low precision.

Machine-learning algorithms can generally model complex class signatures, accept a variety of input predictor data, and do not make assumptions about the data distribution. A wide range of algorithms are being used, such as SVMs, single DTs, RFs, boosted DTs, an AIS based, and k-nearest neighbour (k-NN). Selecting a machine-learning classifier for a particular task is challenging, not only because of the wide range of available machine-learning methods but also because the literature appears contradictory, making it difficult to generalize the relative classification accuracy of individual machine-learning algorithms [6,8,9].

In recent years, a new intelligence theory – AISs has also been applied to classify remote-sensing images. The AIS is inspired by its natural counterparts and has exhibited many benefits over traditional classifiers [7,10,11]. Compared to the conventional statistical classifier, the AIS classifier has the capacity for self-learning and robustness. AIS are data-driven self-adaptive methods that can adjust themselves to the data without any explicit specification of functional or distributional form for the underlying model. Also, AIS are nonlinear models, which makes them flexible in modelling complex real-world relationships. Large-size image data processing for pattern recognition and classification will require high computing facilities [12]. However, AIS algorithms exhibit high efficiency and have potential advantages. These algorithms have demonstrated their strength for efficient and accurate classification by showing good recognition, reinforced learning, feature extraction, memory, diversity and robustness, etc. In addition, they have strong capabilities in pattern recognition. Experimental results by several researchers suggest that these artificial immune classifiers for remote sensing imagery can yield better results than traditional classification algorithms, such as the ML classifier.

5.1.1   The immune system

The human immune system is a complex system of cells, molecules and organs representing an identification mechanism capable of perceiving and combating dysfunction from our cells and the action of exogenous infectious microorganisms [13–15]. Any molecule that the adaptive immune system can recognize is known as an antigen (Ag). The essential component of the immune system is the lymphocytes or the white blood cells. Lymphocytes exist in two forms, B cells and T cells. These two types of cells are rather similar but differ in relation to how they recognize antigens, and by their functional roles, B-cells are capable of recognizing antigens free in solution, while T cells require antigens to be presented by other accessory cells. Immune network theory was proposed by Jerne in 1974 [16] in an attempt to explain the memory-retention and learning capabilities exhibited by the immune system. Unlike the clonal selection principle, the immune network theory hypo-thesizes that the immune system maintains a regulated network of cells and molecules that maintain interactions between not only an antibody and an antigen but also the antibodies themselves. If an antigen is recognized by an antibody ab1, then ab1 may be recognized by ab_2, and, in turn, ab_2 may be recognized by ab_3, forming a network of antibody interaction (Figure 5.1). Recognition among anti-bodies would elicit a negative response and result in the tolerance and suppression of antibodies. In this way, excessively similar antibodies of the same types will be suppressed to guarantee the appropriate number of antibodies. As a result, the immune system will achieve the final state of stability, where these highly adapted antibodies are transformed into long-term memory antibodies. This ensures that memory antibodies can be uniformly distributed in an antigen space. In this way, although there are a relatively small number of antibodies in the immune system, they can cover the entire antigen space and recognize all antigens These immune network principles have been utilized in this study.

Classifiers based on AIS are based on the assumption that the training set constitutes the initial antibodies' population of the system and a suppression mechanism that tries to reduce this training set into a smaller subset [5]. This subset is supposed to contain the most significant samples without losing many capabilities of generalization. Experimental results suggest that these artificial immune classifiers for remote sensing imagery can yield better results than

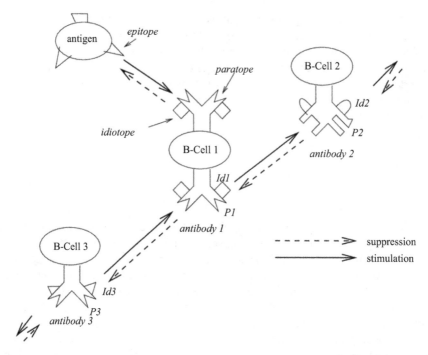

Figure 5.1   Immune network principle (modified after Jerne 1974)

traditional classification algorithms, such as the ML classifier. However, they often require additional user-defined parameters to update the antibody and memory cell population. Researchers modified classical AIN and proposed multiple-valued immune network classifier (MVINC) for the classification of multispectral remote sensing images [17,18]. MVINC builds up an immune network composed of three layers – the antigen, the B cell and the T cell layer – by analogy with the interaction between B cells and T cells in the immune system.

When inputting an antigen, MVINC produces the weighing vector to describe the stimulation level of an input antigen pattern to different T cells. Subsequently, by applying a function modified in real-time, MVINC produces the feedback vector from the T cell to the B cell layer to update the network, called the memory pattern or multiple-valued memory pattern. MVINC constantly trains the immune network to the samples of regions of interest (ROIs) using the aforementioned process until the maximum recorded error is within the tolerance threshold ρ. During the classification process, MVINC learns to classify inputs based on a multiple set of characteristics from 0 to $(m-1)$, indicating the extent to which each one is present – where m represents the total number of characteristics. For the classification of multispectral imagery, m can be replaced with the maximum grayscale value of the image (e.g., $m = 255$). The number of the B cell layers is equal to the number of the bands, while the number of the T cell layers corresponds to the number of the classes, implying that each class has only one T cell [19,20].

5.1.2   Classification based on the AIS

As AIS emerged in the1990s, classification has been an important application area of AIS. Classification systems based on AIS have attractive features inherited from biological immune system [13,21,22]. Classifiers based on AIS are almost on the assumption that the training set constitutes the initial antibodies' population of the system and a suppression mechanism that tries to reduce this training set into a smaller subset. This subset is supposed to contain the most significative samples, without losing much capability of generalization. AIS is concerned with developing a set of memory cells that give a representation of the training data [23]. When an antigen invades, first, AIS evolves a candidate memory cell through the process of clone, mutation and resource competition and then determines whether this candidate cell should be added to the pool of memory cells or not. Classification algorithms based on AIS develop a set of memory cells that give a representation of the training data. When an antigen invades in, the algorithm evolves a candidate memory cell through the process of clone, mutation and resource competition, and then determines whether this candidate cell should be added to the pool of memory cells or not.

5.2   Data used and study area

For urban feature extraction and classification, high-resolution data is required, or we can say that higher classification accuracy can be achieved through it. Worldview-2 data is one of the high-resolution data which is best suited for our study. WorldView-2 satellite collects very high-resolution data in 8 bands. It has two pushbroom sensors that acquire an image in panchromatic and multispectral modes. The panchromatic has a spatial resolution of 0.46m, which is resampled to 0.50m. The multispectral has a spatial resolution of 1.85m, which is resampled to 2.0m. The panchromatic image is a single-band image, while the multispectral is 8 bands image with a radiometric resolution of 11-bit. The detailed explanation of sensors onboard Worldview-2 is shown in Table 5.1.

Table 5.1   Data characteristics

Band	Spectral range	Spatial resolution	Radiometric resolution
Panchromatic	0.450–0.800 μm	0.50m	11-bit(0 values) to 2047 grey scale
Coastal	0.400–0.450 μm	2.0m	11-bit(0 values) to 2047 grey scale
Blue	0.450–0.510 μm	2.0m	11-bit(0 values) to 2047 grey scale
Green	0.510–0.580 μm	2.0m	11-bit(0 values) to 2047 grey scale
Yellow	0.585–0.625 μm	2.0m	11-bit(0 values) to 2047 grey scale
Red	0.630–0.690 μm	2.0m	11-bit(0 values) to 2047 grey scale
Red Edge	0.705–0.745 μm	2.0m	11-bit(0 values) to 2047 grey scale
Near Infrared1	0.770–0.895 μm	2.0m	11-bit(0 values) to 2047 grey scale
Near Infrared2	0.860–1.040 μm	2.0m	11-bit(0 values) to 2047 grey scale

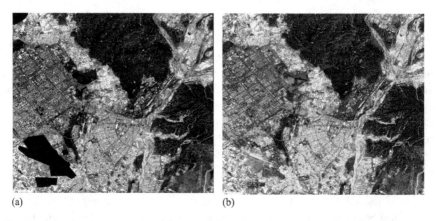

(a)   (b)

Figure 5.2   (a) High-resolution image of study area. (b) Multispectral image of study area.

The study area should be best suited to the adopted methodology for achieving the objectives of our research work, or simply it should be helpful for implementing our methodology to achieve the desired goal (or it will work in an opposite manner also as we opt for that methodology which is best suited to our predefined study area). So, the selection of a study area is a very important part of any research work. Keeping all these things in mind, we take Chandigarh City as a study area for our research work. As Chandigarh is a properly planned city in India, urban feature classification can be very easy due to the numerous urban features with proper spacing.

The study is carried out in a part of Chandigarh city, India. The study area is geographically located between 76°45'32.44" E to 76°56'11.82" E longitudes and 30°47' 52.77" N to 30°38'38.4" N latitudes. The average height of the underlying terrain is 330.77m above mean sea level. Chandigarh is located on the foothills of the Himalayas. The study area consists of a dense urban area including buildings, schools, hospitals, industries, vegetation cover surrounding the school and open spaces, bare land, and road networks. The WorldView-2 images of the study area are shown in Figure 5.2(a) and (b).

5.3   Experimental approach

The remote sensing image classification procedure involves two steps. The first stage is the training of the system with a set of sample data. Generally, sample data is obtained by selecting the ROI. A clonal selection based supervised classification algorithm which takes into account not only spectral features but also geographical features and image features. The proposed algorithm searches the best cluster centres for various types of training samples by the improved clonal selection algorithm. After the training is complete, the remote sensing images are subjected to classification. The training is done on a set of sample images. The sample images are obtained by selecting a ROI. The overall methodology is shown in Figure 5.3. The training procedure is as follows.

5.3.1 Initialization

Available Ab repertoire that can be decomposed into several different subsets. Let Ab{m} represent the set of memory cells. Ab{r} represent the set of remaining Ab. Ab = Ab{m} + Ab{r} (r + m = N). This is done by randomly choosing training

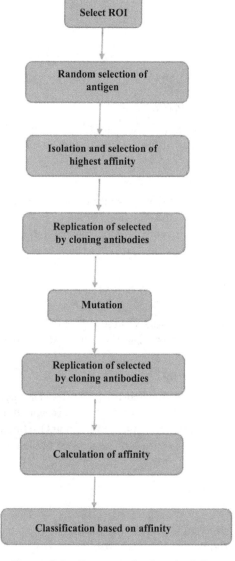

Figure 5.3 Framework of methodology

antigens to be added to the set of memory cells Ab{m} and to the set of Ab{r}. For each antigen (Ag) in the training set perform, the following steps.

5.3.2   *Randomly choose an antigen*

Ag_j in ROI and present it to all Ab's. Determine the vector aff_j that contains the affinity of Ag_j to all the N Ab's in Ab. For the current investigation, Euclidean distance d_j is the primary metric of affinity. The Affinity aff_j is defined as in (5.1) below:

$$d = \sqrt{\sum_{i=1}^{bm} (x_i - y_i)^2} \tag{5.1}$$

$$aff_i = - d_j$$

5.3.3   *Select the* n *highest affinity*

Ab's from Ab to compose a new set $Ab^j{}_{\{n\}}$ of high-affinity Ab's in relation to Ag_j and In Ab $_{\{m\}}$ find the highest affinity memory cell, m match c.

5.3.4   *Clone the* n *selected Ab's*

Based on their antigenic affinities, generating the clone set C^J. The higher the antigenic affinity, the higher the number of clones generated for each of the n selected Ab's. The total number of clones generated Nc is defined in (5.2) as follows:

$$N = \sum_{i=1}^{n} round \left(\frac{\beta \cdot N}{i} \right) \tag{5.2}$$

where β is the a multiplying factor, N is the total number of Ab's round, and $(\cdot)$ is the operator that rounds its argument toward the closest integer.

5.3.5   *Allow each Ab's in clone set*

C^j the opportunity to produce mutated offspring C^{j*}. The higher the affinity, the smaller the mutation rate. Where mutate procedure and function mutate(x) are defined in the equation below. The function Irandom() returns a random value in the range [0,1] and Lrandom returns a random value in the range [−1,1]. Function $\Delta(t, y)$ is defined in (5.3) as follows:

$$\Delta(t,y) = y \left(1 - r^{\left(1 - \frac{t}{T}\right)^{\lambda}} \right) \tag{5.3}$$

where t is the iteration number; T is the maximum of iteration number; r is a random value in the range [0,1]; λ is a parameter to decide the nonconforming degree.

5.3.6 Calculate the affinity aff * j

Calculate the affinity aff $*_j$ of the matured clones C^{j*} in relation to antigen Agj.

5.3.7 Select the highest affinity

Select the highest affinity from the set of C^{j*} in relation to Agj as the candidate memory cell, mc$_{candidate}$, to enter the set of memory antibodies Ab$_{\{m\}}$.

5.3.8 Decide

Whether the mc$_{candidate}$ replaces mc$_{match}$ that was previously identified. If mc$_{candidate}$ has more affinity by the training antigen, Ag. The candidate memory cell is added to the set of memory cells Ab $_{\{m\}}$ b and replaced with mc$_{match}$.

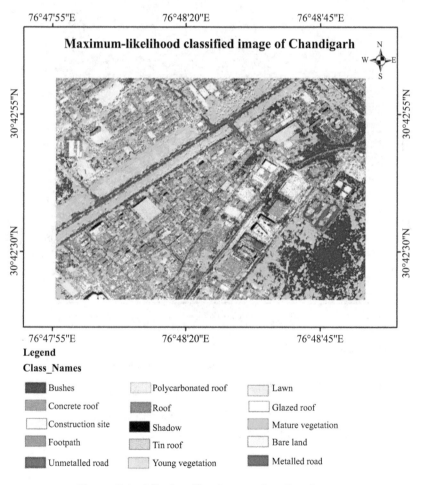

Figure 5.4 ML classification results of study area

5.3.9 Replace

Replace the d lowest affinity Ab's from Ab $_{\{r\}}$.

5.3.10 A stopping criterion

A stopping criterion is calculated at this point. It is met if the average affinity for Ab's is above a threshold value. If the stopping criterion is met, then training on

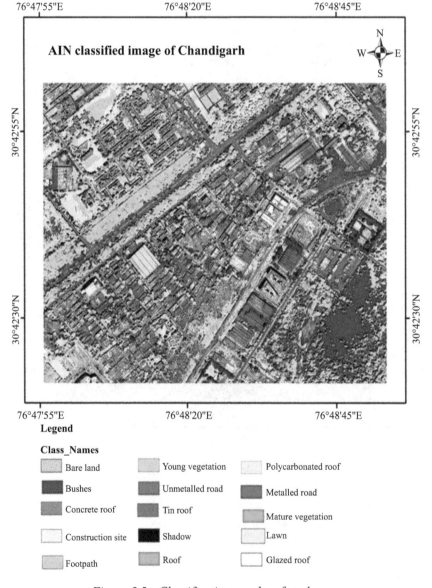

Figure 5.5 Classification results of study area

this one antigen stops. If the stopping criterion has not been met, repeat, beginning at step 3.

After training is done, the evolved memory cells are available for use in classification. Each memory cell is presented with a data item. By calculating the closeness or affinity between the memory cell and image data, the image is classified into the class that has the maximum closeness. Based on the training data, image was classified into 15 urban landcover classes such as different types of roads, roof types and vegetation. The classified results for the maximum-likelihood classifier (Figure 5.4) and AIN (Figure 5.5) are shown below.

5.4   Result

To validate the classification results, an accuracy assessment has been carried out on both classified results. The accuracy was calculated for individual classes to properly compare the classification algorithms. Table 5.2 gives the class-wise achieved accuracy for both classification algorithms.

It was observed that for most of the classes, an improvement in overall accuracy and kappa statistics was observed by using the approach based on an AIN. Table 5.2 shows that the AIN approach produces better classification results than the ML method. Overall kappa statistics for MLC was found to be 0.72, and for AIN, it was calculated to be 0.86. As shown in Table 5.2, the AIS approach improved overall classification accuracy for each class; classes such as lawn, footpath, glazed roof, polycarbonate roof, and concrete roof exhibit the most significant improvement in accuracy, followed by the concrete roof, construction site, road, tin roof, etc. For a few classes, both classifiers exhibited similar accuracy levels, such as young and mature vegetation and bushes. The accuracy levels

Table 5.2   Performance analysis

Classes	MLC	AIN classifier
Young vegetation	1.00	1.00
Mature vegetation	0.95	0.97
Shadow	1.00	1.00
Metalled road	1.00	0.58
Polycarbonate roof	0.11	0.66
Lawn	0.33	0.61
Unmetalled road	0.57	0.40
Bare land	0.46	1.00
Roof	0.52	0.47
Footpath	0.20	0.74
Tin roof	0.32	0.47
Glazed roof	0.23	1.00
Construction site	0.20	0.37
Bushes	0.92	0.85
Concrete roof	0.47	0.55
Overall kappa	0.56	0.72

decreased for metalled and unmetalled roads. This is because the ML approach works well only when the underlying assumptions are satisfied and poor performance may be obtained if the actual probability density functions are different from those assumed by the model. At the same time, AIS are nonlinear models, which make them flexible in modelling real-world complex relationship.

5.5   Conclusion

In this paper, we synthesise the advantages of the AIS and proposed a new remote sensing image classification algorithm using clonal selection algorithm which is a basis of the immune system. A quantitative comparison between the conventional ML statistical classifier and our algorithm was demonstrated that the ML statistical classifier is less capable of discriminating roof types and urban features than AIS classifier. The results also show concurrence with field observations and classification map derived through visually interpreted reference. Experimental results show that the proposed classification algorithm has high classification precision. It is a good and efficient classification algorithm and can be applied to remote sensing image classification. AIS will provide an alternative approach for accurate pattern recognition in remote sensing data. It will not only be an effective learning algorithm and classify multi-remote sensing images but also a very competent classifier for processing high volumes of data e.g. in hyperspectral images. As a future scope of the work, it is planned to investigate the approach for decreasing the number of unclassified antigens which are not recognized by any artificial antibody in the trained network. In addition, it is also planned to enhance the classifiers by considering feature selection or extraction using other AIS models in high-dimensional feature space.

References

[1]   Watkins, A. and Timmis, J. (2004). Artificial immune recognition system (AIRS): an immune-inspired supervised learning algorithm. *Genetic Programming and Evolvable Machine*, 5, pp. 291–317.

[2]   Zhang, H., Yang, H., and Guan, C. (2013). Bayesian learning for spatial filtering in an EEG-based brain–computer interface. *IEEE Transactions on Neural Networks and Learning Systems*, 24, pp. 1049–1060. doi: 10.1109/TNNLS.2013.2249087.

[3]   Zhong, Y., Zhang, L., Huang, B., *et al.* (2006). An unsupervised artificial immune classifier for multi/hyperspectral remote sensing imagery. *IEEE Transactions on Geoscience and Remote Sensing*, 44(2), pp. 420–431.

[4]   Liu, J.Y. Liu, M.L., Zhuang, D.F., Zhang, Z.X., and Deng, X.Z. (2003). Study on spatial pattern of land-use change in China during 1995–2000. *Science in China Series D*, 46, pp. 373–384.

[5]   Andrew, S., Alex, F., and Jon, T. (2003). AISEC: an artificial immune system for e-mail classification. In: *The 2003 Congress on Evolutionary Computation*, IEEE, 1, pp. 131–138.

[6] Greensmith, J. and Cayzer, S. (2002). An artificial immune system approach to semantic document classification. In: *Proceedings of ICARIS* 2002, pp. 136–146.

[7] Jerome, H. and Carter, M. (2000). The immune system as a model for pattern recognition and classification. *Journal of the American Medical Informatics Association*, 7(1), pp. 28–41.

[8] Mayaud, L., Cabanilles, S., Van Langhenhove, A., *et al.* (2016). Brain-computer interface for the communication of acute patients: a feasibility study and a randomized controlled trial comparing performance with healthy participants and a traditional assistive device. *Brain Computer Interfaces*, 3, pp. 197–215, doi:10.1080/2326263X.2016.1254403.

[9] Nasir, R., Javaid, I., Fahad, M., *et al.* (2018). Artificial immune system – negative selection classification algorithm (NSCA) for four class electro-encephalogram (EEG) signals. *Frontiers in Human Neuroscience*, 12, p. 439, doi:10.3389/fnhum.2018.00439.

[10] De Castro, L. (2000). An evolutionary immune network for data clustering. In: *Proceedings of the IEEE SBRN 2000*, Brazil, pp. 84–89.

[11] Zhang, L., Zhong, Y., Huang, B., and Li, P. (2007). A resource limited artificial immune system algorithm for supervised classification of multi/hyper-spectral remote sensing imagery. *International Journal of Remote Sensing*, 28(7), pp. 1665–1686, 2007.

[12] McCoy, D. and Devarajan V., 1997. Artificial immune systems and aerial image segmentation. In: *Systems, Mans, and Cybernetics and Stimulation, 1997 IEEE International Conference on System*, 1, pp. 867–872.

[13] Adriane, S. and Jose, M. (2007). Artificial immune systems for classification of petroleum well drilling operations. In: *ICARIS 2007, LNCS 4628*, pp. 47–58.

[14] George, B. and Tiago, B. (2005). Adaptive radius immune algorithm for data clustering. In: *ICARIS 2005, LNCS 3627*, pp. 290–303.

[15] Hart, E., and Timmis, J. (2008). Applications of artificial immune systems: the past, the present and the future. *Journal of Soft Computing*, 8(1), pp. 191–201.

[16] Jerne, N.K. (1974). Towards a network theory of the immune system. *Annual Review of Immunology*, 125c, pp. 51–60.

[17] Meisheri, H., Ramrao, N., and Mitra, S. (2018). Multiclass common spatial pattern for EEG-based brain-computer interface with adaptive learning classifier. arXiv [preprint]:1802.09046.

[18] Wang, J., Feng, Z., and Lu, N. (2017). Feature extraction by common spatial pattern in the frequency domain for motor imagery tasks classification. In: *2017 29th Chinese Control and Decision Conference (CCDC)*, IEEE, pp. 5883–5888.

[19] Zhong, Y., Zhang, L., Gong, J., and Li, P. (2007). A supervised artificial immune classifier for remote-sensing imagery. *IEEE Transactions on Geoscience and Remote Sensing*, 45(12), pp. 3957–3966.

[20] Zhang, L., Zhong, Y., and LI, P. (2004). Application of artificial immune systems in remote sensing image classification. In: *The International Archives of Photogrammetry, Remote Sensing and Spatial Information Sciences*.

[21] Campelo, F., Guimaraes, F.G., and Igarashi, H. (2008). Multiobjective optimization using compromise programming and an immune algorithm. *IEEE Transactions on Magnetics*, 44(6), pp. 982–985.

[22] Grazziela. F., Nelson, E., Helio, J.C.B., *et al.* (2007). The SUPRAIC algorithm: a suppression immune-based mechanism to find a representative training set in data classification tasks. In: *ICARIS 2007*, LNCS 4628, pp. 50–70.

[23] Yin, G. (2003). The multi-sensor fusion: image registration using artificial immune algorithm. In: *Soft Computing Techniques in Instrument, Measurement and Related Applications* 2003, pp. 32–36.

Chapter 6

Detection and segmentation of aircrafts in UAV images with a deep learning-based approach

Hina Pande[1], Poonam Seth Tiwari[1], Parul Dhingra[1] and Shefali Agarwal[1]

Advancements in unmanned aerial vehicles' (UAVs) technology have enabled the acquisition of images of a geographical area with higher spatial resolutions as compared to images acquired by satellites. Detection and segmentation of objects in such ultrahigh-spatial-resolution (UHSR) images possess the potential of effectively facilitating several applications of remote sensing (RS) such as airport surveillance, urban studies, and road traffic monitoring crop monitoring. Investigating these images for target extraction tasks turns out to be quite challenging, in the terms of the involved computation complexities, to their high spatial resolutions and information content. Due to the development of several deep learning (DL) algorithms and advanced computing tools, there exists a possibility of harnessing this information for computer vision (CV) tasks. Manual surveillance of airports or similar areas, and manual annotation of images are cost intensive and prone to human-induced errors. Therefore, there exists a substantial requirement of automating the task of keeping track of the airplanes parked in the premises of airports for civil and military services. With this paper, we propose a framework for detecting and segmenting such airplanes in UHSR images with supervised machine learning algorithms. To detect the target i.e. airplanes, MobileNets-deep neural network is trained, whereas to segment the target, U-Net-convolutional neural network is trained with our dataset. Further, the performance analysis of the trained deep neural networks is presented. The UHSR image dataset utilized in this research work is an airport-dataset provided by SenseFly. Data is acquired by eBee classic drones, flying at a height of 393.7 ft., which provides 2D-RGB images with ground resolution of 3.14 cm/px.

6.1 Introduction

UAVs [1] are utilized in civil and military arenas for several purposes such as surveillance, security, recreational, educational, rescuing, and monitoring. Due to

[1]Indian Institute of Remote Sensing, ISRO, India

their low maintenance cost, undemanding installations, and ability to manoeuvre over a geographical region with high mobility and reliability they are better suited for such purposes as compared to satellites. However, weather plays a crucial role in their functionality, as bad weather conditions can adversely affect their manoeuvring capabilities. UAVs equipped with advanced high-resolution cameras readily provide images and videos of a geographical area with continuity, reliability and fine details. These images or videos can be further analyzed and processed to extract meaningful information from them for various applications. Over the past few years, UAVs are employed to conduct search and rescue operations in sea [2], sense the temperature of streams by using thermal sensors [3], monitor crops and droughts, transport goods, inspect construction sites, and various other active and passive RS applications [4].

The images acquired by aerial missions are subjected to various image processing steps [1] to increase their readability and quality. First, the initial estimates of orientation and position of each image are acquired by the log files. To re-establish the true orientation and position of the images acquired by UAVs, aerial triangulation is implemented. With this step, many automated tie points are generated for conjugate points corresponding to multiple images. The automated tie points are used to optimize the image orientation and position with bundle block adjustment. Further, a digital surface model is created with oriented images. The features in multiple image pairs are matched which generate dense point cloud. Subsequently, a digital terrain model is generated, and, to remove distortion in images, orthorectification process is implemented. The images are then combined into a mosaic to produce seamless images of geographical area of interest. RS image is a key resource widely used in civil and military applications. Automated image interpretation and object detection is one of the most important tasks in RS. The availability of high resolution images and richer information content of this data has captured the attention of academia and industry.

Aircraft is an important means of transportation in civil and military applications and hence is one of the most important targets in the field of object detection. The accurate detection of aircraft has crucial significance and military value. Therefore, aircraft detection from RS images has become the focus of attention. The high resolution images have the ability to accommodate abundant and finer information about terrains, and therefore, possess capability of discerning objects distinctly. This has led to research proliferation towards object detection in the field of RS. The traditional object detection techniques [5] include selecting a desired area in the image, extracting the features in the desired area, and lastly, for classification using the training classifier. The few examples [6] of traditional object detection are feature descriptors like SURF, BRIEF, SIFT etc. for object detection, and machine learning algorithms like SVM, K-nearest neighbour, etc. for predictions. However, these techniques lack robustness and adaptability, and thus, require rigorous tuning of thresholds and parameters for different environments. DL paves the way for increasing the robustness of detection algorithms, as they have the capability to perform better in the environments where brightness, SNR, and backgrounds in an image differ, and detect wider range of objects in an image.

Various DL algorithms like R-CNN, Fast R-CNN, YOLO, Faster R-CNN, SSD, R-FCN, etc. have been developed for various object detection tasks.

Traditionally convolutional networks were used for classification tasks [7]. However, there are several segmentation applications that require assignment of class labels to each pixel. Ciresan *et al.* [8] proposed a network for such requirements, where the local region around the pixel was used to predict the class label. This algorithm was quite slow as it had to run separately for each local region/ patch, and there was a trade-off between use of context in terms of size of patches and accuracy of localization. To overcome these limitations networks were developed where features from multiple layers were taken into account for classifier output. Further, Olaf Ronneberger *et al.* [7] developed a more advanced architecture where even with few training images precise segmentations could be achieved. Pathak *et al.* [14] applied DL for object detection. Alganci *et al.* [15] compared various DL approaches such as Faster R-CNN, Single Shot Multi-box Detector (SSD), and You Only Look Once-v3 (YOLO-v3) for airplane detection from very high-resolution satellite images. They concluded that Faster R-CNN architecture provided the highest accuracy according to the F1 scores, average precision (AP) metrics and visual inspection of the results. Zhaoa *et al.* [30] proposed a heterogeneous model to transfer CNNs to remote-sensing scene classification to correct input feature differences between target and source datasets. Ji *et al.* [31] detected aircraft in high spatial resolution RS images by combining multi-angle features and majority voting CNN.

The study aims to develop a supervised learning framework for detecting airplanes in UHSR images acquired with UAV's using MobileNet-deep neural network. Since the target of interest is likely to have multiple orientations in the image, a multi-angle feature extraction is enabled. The airport images are manually labelled and segmented using U-Net architecture. Performance analysis of deep neural networks is assessed in the study. The paper is organized as follows. Section 6.1 discusses the basic technical concepts underlying our research work. The method for detection and segmentation of airplanes in UHSR images is presented in Section 6.3. Section 6.4 discusses training and testing process, limitations, and objective analysis of the trained models. Section 6.5 states conclusions.

6.2   Background

UAV's has recently become popular across the fields of CV and RS due to their comprehensive and flexible data acquisition, Inspired by recent success of DL, many advanced object detection and tracking approaches have been widely applied to various UAV-related tasks, such as environmental monitoring, precision agriculture, and traffic management.

The following section explains characteristics of high resolution images, and further, provides the background of the neural networks and convolutional neural networks (CNN) specifically for automated object detection and segmentation.

6.2.1   Digital images and spatial resolution

Digital image analysis and processing enables formulating techniques to remove noise from images, increase their interpretability, extract desired object, and compress them for storage or transmission purposes. Images can be either in digital or analogue format. 2D-Digital images are signals, say Im(x,y), where x and y are two independent variables (spatial coordinates). The basic constituent of a digital image is a pixel. Pixels are picture elements that are square in shape. Digital images are a rectangular array of pixels [9]. For RS applications, the sensors mounted on a platform capture the energy emitted or reflected by objects present in a geographical area. The sensors can be mounted on satellites, airplanes, or UAVs according to the desired application. The value of each pixel is directly proportional to the intensity of the light captured and recorded by the optical sensors at a given point. A grayscale image, where each pixel can attain a value in the range of 0–255, can be represented by a single 2D array, whereas an RGB image with three channels is represented by a collection of three 2D arrays, each for the red, green, and blue channel. Pixels are the physical points containing a digitized value recorded by optical sensors. Digital images can be referenced with rows and columns. Digital images are stored in various image file formats [10] such as bitmap (BMP), tagged image file format (TIFF), joint photographic expert graphic (JPEG), and portable network graphics (PNG). There are four types of resolutions that define the characteristics of a digital image, namely, spectral, spatial, radiometric, and temporal. The spatial resolution [10] of an image corresponds to the actual area in the scene represented by a single pixel in an image when a sensor performs imaging with the instantaneous field of view (IFOV). It is a measure of the smallest object in a scene that can be discerned by the optical sensor sensing over a geographical area. The high resolution images allow us to differentiate objects that are closer to each other. The study uses 2D-digital images that are acquired by capturing the reflected sunlight energy in the visible region of the electromagnetic spectrum. These images have UHSR of 3.14 cm.

6.2.2   Neural networks

The human brain is a highly non-linear data processing system, where complex computations are performed extremely fast. An artificial neural network tries to mathematically model the functioning of a brain for performing such non-linear computational tasks. The basic constituent of an artificial neural network is artificial neuron [11]. A neural network is developed by interconnecting these neurons. The three basic elements of an artificial neuron are as follows. First, the connecting links between the inputs and the neuron. With each link, there is a synaptic weight associated to it. Synaptic weights of artificial neurons can obtain negative as well as positive values. The input to a neuron is the summation of weighted inputs, with a bias added to it. Second, an adder to add weighted inputs and bias. The mathematical operation carried out at adder yields output v_k:

$$v_k = \sum_{j=1}^{m} w_{kj}x_j + b_k \qquad (6.1)$$

At kth neuron, x_j is input signal at jth synapse which is multiplied by synaptic weight w_{kj}, and b_k is bias. Third, an activation function to limit the value of the output from the neuron. They are also called as squashing functions, as they squash the values of outputs to permissible finite values. The mathematical notation for an activation function (.) yielding output y_k is:

$$y_k = \varphi(v_k) \qquad (6.2)$$

The neural network consists of one or more layers comprising of neurons. A single-layered neural network is a network where inputs are fed directly to the output layer. In multi-layer networks, each neuron of a layer is connected with all the nodes of the input layer. The neurons within a layer are not connected. The feed-forward networks are the networks in which the signals are forwarded from one layer to the next without any feedback loop. There can be several layers between the source layer and the output layer. The layers in between the input layer and output layers are called hidden layers. The hidden layers enable the extraction of high-order statistics from input signals. The recurrent neural networks are the networks that contain at least one feedback loop. The recurrent networks are designed such that the output from every neuron in a layer is fed back to the network as input to all the neurons. The artificial neural network can be trained with supervised, unsupervised, or semi-supervised learning algorithms. In supervised learning, both the input signals and their corresponding desired outputs are utilized for training the network, thus we require labelled data. The input signal is fed to the network, and the loss is calculated by taking into account the predicted outcome and ideal expected outcome. The unsupervised learning algorithms train the neural networks with unlabelled input signals. The semi-supervised way of learning takes into account both labelled and unlabelled training input signals. The applications of artificial neural networks for automating various tasks are automated driver assisting systems, speech recognition, handwriting recognition, etc. We have trained the neural network with supervised learning algorithms for automatic detection and segmentation of the target, i.e. airplanes.

6.2.3 CNNs

CNN are widely used neural networks for extracting information from 2D-image data, where inputs are grid-structured, and there are spatial dependencies within the local regions [12]. The pixels in the neighbourhood of an individual pixel often have similar values; hence, image data exhibits strong spatial dependencies, which makes it highly suitable for CNNs. The CNNs can be used for spatial, temporal, and spatiotemporal input data. The image data exhibits translation invariance, where an object has the same interpretation irrespective of its location in the image. In CNNs, similar feature values are created from local regions that have a similar pattern. The basic operation executed in CNNs is mathematical convolution. A convolution operation is a sliding dot-product carried out between the convolution filters and grid-structured inputs. The operation is beneficial for data that exhibits a high level of spatial locality. CNNs are the neural networks in which at least one layer is the convolution layer. There can be one or multiple convolution layers in a neural network. As every feature value in the current layer is dependent on the small local region in the previous layer, the spatial relationships get inherited from one layer to the next layer. A three-dimensional grid structure with height, width, and depth define each convolution layer of a CNN. The depth refers to the number of feature maps in a convolutional layer. The basic building blocks of a typical feed-forward convolutional neural network are convolution layer, pooling layer, rectified linear unit (ReLU) layer, fully connected layer, and loss layer. The convolution layer overlaps the kernel at every location in an image and performs a sliding dot product. The pooling layer basically performs the down sampling of the feature maps in a non-linear manner. Max pooling is one most commonly used non-linear function in the pooling layers. A new feature map is produced as the pooling layer acts independently on every depth slice of the feature map. The input image is partitioned into non-overlapping regions in a feature map, and the pooling function obtains the maximum value in the particular region to generate a new feature map. The pooling layer reduces the size of the feature map and the parameters required to train the network; hence, the computational complexities within the convolutional neural network are reduced. The commonly used activation function in CNNs is ReLU activation functions. The ReLU function squashes the negative values to zero and, therefore, does not permit negative values to propagate in the network. The dimensions of a layer remain the same when an activation function is applied as it only maps the values in the feature map corresponding to the activation function. After the implementation of convolution and max-pooling layers, lastly, the outputs are generated by implementing a fully connected neural layer. The loss layer which is the final layer of the convolutional neural network determines the deviation between the expected ideal outcome and the predicted outcome. Softmax loss and sigmoid cross-entropy loss are examples of such loss functions in the loss layer. The CNNs are used to perform object detection, classification, and segmentation tasks in CV. We have implemented CNNs to automate the task of target extraction from UHSR images.

6.3   Methodology

6.3.1   Dataset

The UHSR image-dataset utilized in the project is captured by two eBee classic drones [13], flying at a height of 393.7 ft. The ground resolution of images is 3.14 cm/px. The data has been acquired over the Le Bourget airport in Paris. The dimension of the images is 4,608 × 3,456 pixels. The 2D-images are captured in the visible spectral range: comprising the red, green, and blue wavelengths. The images contain one or multiple parked-airplanes, along with several other objects like buildings, runways, automobiles, etc. Figure 6.1 shows images from airport dataset.

The following section describes the experimental approach adopted for extracting airplanes from the UHSR images in an automated way. Figure 6.2 depicts the broad methodology followed for the study.

(a)

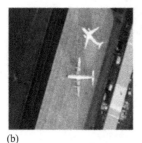

(b)

Figure 6.1 (a) SenseFly eBee drone [1]. (b) Image samples of airport area.

6.3.2 Object detection

Object detection implies that we aim towards enclosing the target object i.e. airplane with rectangular bounding boxes. The deep neural frameworks perform better for this category of application than shallow networks [14,15]. Our training set consists of 13 RGB images. The images are annotated with LabelImg where annotations are stored in PASCAL VOC format. The trained network is tested with nine test images.

6.3.2.1 Data pre-processing

The dimension of images is reduced to 800 × 600 pixels from 4,608 × 3,456 pixels to reduce the computational complexities and time required for training the deep neural network.

6.3.2.2 Image annotation using LabelImg

The processed images are manually annotated as shown in Figure 6.3 with LabelImg [16], version 1.8.0, open-source software for graphical image annotations. It generates the annotation files in XML (Extensible Markup Language)-.xml format. The XML file saves the name of the image, size (800,600) and depth (3) of

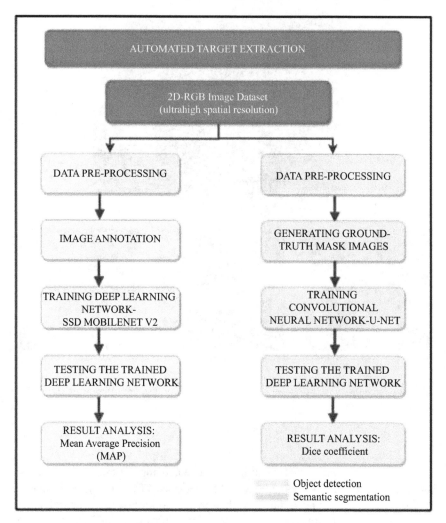

Figure 6.2   Methodology for automated target extraction

the image, name of the object annotated (airplane) and location of the manually annotated bounding boxes in the image. For training, the neural network 13 images are annotated manually which generated 13 .xml files.

Further, these 13 individual .xml annotation files are combined and converted to Comma Separated Values (CSV) .csv file. The .csv file and image data are converted and stored in TensorFlow Record (TFRecord) format. The TFRecord format stores the data in binary format and significantly reduces the training time, and occupies less space in the storage disk. The training data in TFRecord file format is fed to the neural network.

Figure 6.3   Image annotation with LabelImg

6.3.2.3   Network architecture: SSD MobileNet-v2

The SSD-MobileNet-v2 [17] DL architecture is implemented to classify and locate the airplanes in digital images. The output of the network generates the bounding boxes around the desired feature and gives the confidence score (CS) for the class encapsulated by the box. In this architecture, MobileNet-v2 model is used to classify features and subsequently, a single shot multibox detector model (SSD) is used to locate the feature with bounding boxes. MobileNet models [18] are lightweight neural network models that are based on depth-wise separable convolutions, which facilitates the reduction in the size of the model. There is significantly lesser number of parameters required in this model, as compared to other advanced DL frameworks like GoogleNet and VGG16 [18] for object detection. The convolutional blocks in Mobilenet-v1 consist of two layers, namely, depth-wise convolution layer and pointwise convolution layer. The depth-wise convolution means that to each channel, a single convolutional filter is applied. Further, pointwise convolution is applied to merge the outputs from the depthwise layer. The main difference between standard convolution and depthwise separable convolution is that the former filters and combines the inputs to generate the output in a single step, whereas the later divides it into two layers, first to filter and second to combine. This division helps in reducing the size of the model and hence, associated computations. The MobileNet-v2 [17] consists of inverted residual structure as its backbone, where the thin bottleneck layer possesses the short connections in between them. The inverted residual with a linear bottleneck layer is given the input which is a low-dimensional representation. It expands it to a high dimension and further, depthwise convolution filtering is performed. Finally, with the help of linear convolution, there is a backward projection to lower-dimension. These layers commendably enable the reduction in memory footprint required during inference. Mobilenet-v2 model comprises of two types of block, the residual block (stride = 1) and the downsizing block (stride = 2). The ReLU6 in each block is the rectified linear unit activation function with maximum output limited to 6. The MobileNet-v2 architecture comprises of, first, a fully convolutional layer having 32 filters, and subsequently 19 residual bottleneck layers. The SSD network [19] incorporates a feed-forward convolutional network. The SSD network is appended as the auxiliary network to the base network MobileNet-v2 architecture. The base network works as a feature extractor. The SSD network performs an object–detection task, where its outputs are bounding boxes and the corresponding CSs of the particular class. It implements non-max suppression as the last step for the detection of the object. The SSD MobileNet-v2 DL architecture is one of the most advanced and lightweight deep neural networks.

6.3.3   Semantic segmentation

Semantic segmentation [20] intends to classify each pixel in an image to its corresponding class/label. In semantic segmentation, pixels of multiple objects belonging to the same class are considered as a single entity. We intend to implement semantic segmentation of the image, where all airplanes belong to the same class – 'airplane'. Let there be m pre-defined labels, such that, label = {label, label2,...labelm}, where j = 1 to m. Let the image consists of k number of pixels, such that pixel = {pixel1, pixel2,..., pixelk}, where i = 1 to k, then we intend to implement an architecture 'S' for semantic segmentation such that for each pixel: pixeli, there is a class: labelj assigned to it [20]. We train the U-Net network with 13 RGB images. The trained network is tested with nine images.

6.3.3.1   Data pre-processing

To reduce the computational complexities while training the neural network, images are resized to a dimension of 256 × 256 and are converted to the grayscale format.

6.3.3.2   Image annotations

The U-Net architecture is a supervised machine learning technique. For every training image, we create its corresponding ground-truth mask image. The masks images are such that the pixels belonging to the target possess a value of '255' and the value of background pixels is '0'. The ground truth mask is generated using Microsoft Paint3D. The dimension of ground truth images is 256 × 256 × 1. Figure 6.4 shows the examples of raw image from the dataset. Figure 6.5 presents their ground truth masks.

Figure 6.4   Raw images from training dataset

Figure 6.5   Ground truth mask images

6.3.3.3   Network architecture: U-Net

The implementation of U-Net architecture for semantic segmentation requires less training data as compared to several other CNNs and provides good segmentation results [7]. The U-Net model comprises of two paths, namely, contraction and expansion path. There are a total of 23 convolutional layers in the architecture. To harness the context information of each pixel, the contracting path extracts the features [21] at various levels. It is performed by sequential implementations of convolutions, activation functions, and max pooling. Subsequently, to increase the resolution of the segmented features, the expanding path, which is symmetric to the contraction path, is implemented. It consists of sequential implementations of up-convolutions and ReLU activation functions. Due to the contraction and expansion nature of the architecture, it is called as a U-Net architecture. To capacitate propagating context information to higher resolution layers, the upsampling network consists of a large number of feature channels. There are no fully connected layers in the entire U-Net architecture. The final output from the expansion path consists of an image where the value of each pixel gives its class. The steps for contracting path [7] are as follows. First, it performs two 3 × 3 convolutions with 64 filters. After each convolution, the outputs are subjected to the ReLU activation function and are downsampled by using a 2 × 2 max-pooling operation with stride 2. With each down sampling step, the number of feature channels is doubled. The architecture for the expansion path is as follows. First, the expansion path upsamples the feature map. Subsequently, to reduce the number of feature channels to half, a 2 × 2 up-convolution is implemented, followed by a concatenation step. Further, two 3 × 3 convolutions are implemented. The outputs from both the convolutions are subjected to the ReLU activation function. The final layer implements 1 × 1 convolutions, which relates a feature vector consisting of 64 components to the required number of labels.

6.4   Model training and results

The section presents steps and parameters required for training the deep neural networks, outputs predicted from the trained network and additionally, the limitations of the network. The networks are implemented on the cloud-computing platform by utilizing the data storage and computational capabilities provided by Microsoft Azure and Google Colaboratory.

6.4.1   Object detection

6.4.1.1   Training

To train the network [22], instructions are implemented in Python programming language. Transfer learning technique is used for training the model. A pre-trained model where the base model is trained with Microsoft's Common Objects in Context-dataset [23] is utilized for transfer learning. This reduces the training time and the required computations and provides initial weights/checkpoints of the model. Further, we train the model to tune and update the weights/checkpoints of the model for our dataset consisting of 13 images and their corresponding annotation files. The optimum steps for training/testing, batch and epoch size are arrived at by a trial and error method. The number of training steps and evaluation steps implemented are 4,500 and 100, respectively. The value for batch size is 12. The model uses sigmoid cross entropy loss function for classification purposes and smooth L1 loss function for localization purpose. The model is trained in TensorFlow version 1.15 environment.

6.4.1.2   Test results

The trained SSD MobileNet-v2 deep neural network model is tested with nine RGB images. The object detected is saved using following parameters: x and y coordinates of centre of bounding box, height and width of bounding box, and CS. Figure 6.6 gives four test cases: A, B, C, and D. Figure 6.7 presents predicted outputs from trained network for test cases.

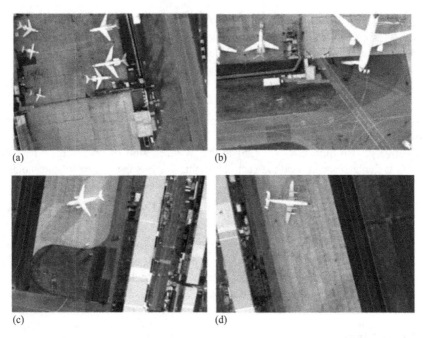

(a)

(b)

(c)

(d)

Figure 6.6   Test cases: RGB images (a), (b), (c), and (d)

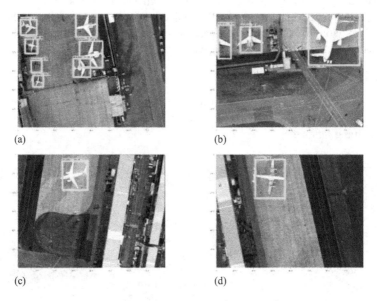

(a)

(b)

(c)

(d)

Figure 6.7   Detection results from trained model

6.4.1.3 Evaluation metrics: mean AP

The CS [24] indicates the probability of presence of an airplane in a bounding box, and accuracy of the box itself. The mathematical equation for calculating CS is:

$$CS = Probability \times IoU \tag{6.3}$$

The intersection over union (IoU) is the ratio of area of overlap between the ground-truth bounding box and the predicted bounding box, and the total area encompassed by both. It calculates the similarity between the predicted bounding box and its respective ground-truth box. AP [25] is the weighted sum of precisions, where the weight is the increase in recall, at each threshold, and *m* is the number of thresholds:

$$AP = \sum_{m=0}^{m=n-1} [Recalls(m) - Recalls(m+1)] \times Precision(m) \tag{6.4}$$

With respect to COCO challenge accuracy metrics [25], we set 10 different IoU thresholds from 0.5 to 0.95 in the steps of 0.05. The accuracy metrics mAP, is calculated by averaging over all the APs of classes detected by object detection model. Our model is trained to detect a single class i.e. airplanes. The mAP = AP value achieved for our trained detection model is 95.9%.

6.4.1.4 Limitations

SSD-MobileNet v2 is trained to enclose the airplane feature in rectangular bounding boxes. When the trained network is tested with nine images, it is observed that if an image consists of an object whose spatial features are similar to that of an airplane, then it captures that object too as the target. Figure 6.8 shows the two limitation cases A and B, where the objects other than airplane feature are enclosed in rectangular bounding boxes.

6.4.2 Semantic segmentation

To train the network, we perform all the computations on Microsoft Azure cloud-computing platform. The instructions to train the model are written in Python programming language.

Figure 6.8 Limitation cases for object detection

6.4.2.1 Training

The network [22] is trained by utilizing neural network libraries provided by Keras API. The training data is stored in portable network graphics (PNG) format. Before initiating the training process, the data augmentation step is carried out, as we are aiming to train the network with a lesser number of images. The data augmentation process helps in minimizing the risk of overfitting while training the neural network. To execute the data augmentation, few of the operations executed are flipping, zooming, shearing, etc. The initial weights of the network are set-up using transfer learning. Further, we train the U-Net with our training data to update the weights with several epochs. The model's weights get updated after every epoch if the loss reduces. The Adam-optimiser is used to update the model's weights after each iteration. The loss function implemented while training the network is binary cross entropy loss [26]:

$$Loss = -\frac{1}{s}\sum_{i=1}^{s} a_i \log \widehat{a}_i + (1 - a_i)\log(1 - \widehat{a}_i) \tag{6.5}$$

where $\widehat{a}_i$ is the ith value in the model output, s is the output size, and a_i is the target value.

6.4.2.2 Test results

The neural network trained is trained with 150 epochs. The loss reduced from 1.5627 to 0.01195 after 150 epochs. Figure 6.9 presents several examples of outputs from trained U-Net neural-network with 150 epochs. The first row consists of original RGB images used for testing the network; the second row presents the corresponding grayscale images which are fed to test the trained neural network, and the third row presents the outputs from the trained UNet neural network.

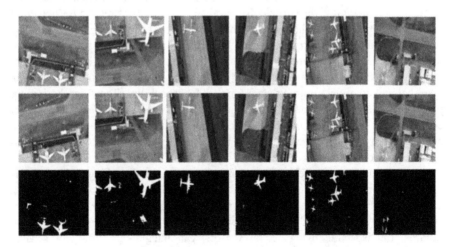

Figure 6.9   Outputs from trained U-Net neural network. First row: original RGB images for testing. Second row: corresponding grayscale images fed to network for testing. Third row: outputs from the trained U-Net neural network.

6.4.2.3   Evaluation metrics: dice coefficient

Dice similarity coefficient (DSC) [27] evaluates the spatial overlap between the ground truth ('M') and result obtained from the trained U-Net ('N') when tested with the test images. The mathematical equation for calculating DSC is:

$$DSC = 2(M \cap N)/(M + N) \tag{6.6}$$

The values of the dice coefficient range in between 0 and 1. The higher value implies higher segmentation accuracy, and high similarity in terms of spatial overlap between ground truth image and output image from U-Net network. The values of DSCs obtained for test set is 0.72.

6.4.2.4   Limitation cases

The training process of U-Net deep neural network includes learning the features of the target object with each iteration. When U-Net model is implemented for our dataset, it is observed that along with the target object, occasionally, it also segments the objects which have a similar shape as that of target object in the same category. Hence, the limitation cases are the instances where test image contains an object (not airplane) that has similar spatial features as an airplane in the spatial domain, and, when tested with the trained U-Net, it segments it as the target object.

6.5   Conclusions and discussion

A framework for automatic detection and segmentation of airplanes in UHSR images captured by UAV has been presented here. To accomplish this task with supervised machine learning algorithms, a state-of-the-art SSD MobileNet-v2 for object detection and U-Net for semantic segmentation has been implemented. The implemented method gives sufficiently good accuracy metrics results for detection and segmentation task in the terms of mAP and dice coefficient respectively. The implemented architectures possess a limitation that objects that are similar in shape to the target, i.e. airplane, are detected or segmented as targets too. The architecture for segmentation of images performs hard classification for each pixel, where each pixel belongs to a class with either 0% or 100% probability. There can be impure pixels in the image, especially at the boundaries of the target, where they might contain both a background object and target object in an image. Research work can be carried out towards developing such neural network architectures that can distinctively specify those pixels as well.

In the near future, development of object detection and tracking in UAV RS is expected and new techniques will emerge to improve these metrics even further. In addition, efficiently processing massive multisource UAV RS data are worth consideration. UAVs equipped with different sensors, e.g., visible, infrared, thermal infrared, multispectral, hyperspectral sensors, can integrate a variety of sensing modalities to make use of their complementary properties, which further realises more robust and accurate object tracking and detection. There is also an urgent need for comprehensive dataset repositories to be used for training. Investigations regarding real-time processing using UAV datasets may also be explored. This study mainly focuses on aircraft detection; however, the method proposed is theoretically applicable to other kinds of targets e.g. ships, vehicles, etc. in RS images. In future, the algorithm could consider more hyperparameters to optimise the network structure and the training process and further improve the accuracy of detection.

References

[1] H. Shakhatreh, A.H. Sawalmeh, A. Al-Fuqaha, *et al.* Unmanned aerial vehicles (UAVs): A survey on civil applications and key research challenges. *IEEE Access*, 7, pp. 48572–48634, 2019, doi:10.1109/ACCESS.2019.2909530.

[2] S. Wang, Y. Han, J. Chen, Z. Zhang, G. Wang, and N. Du. A deep-learning-based sea search and rescue algorithm by UAV remote sensing. In: *2018 IEEE CSAA Guidance, Navigation and Control Conference (CGNCC)*, pp. 1–5, 2018.

[3] A.M. Jensen, B.T. Neilson, M. McKee, and Y. Chen. Thermal remote sensing with an autonomous unmanned aerial remote sensing platform for surface stream temperatures. In: *2012 IEEE International Geoscience and Remote Sensing Symposium*, pp. 5049–5052, 2012.

[4] H. Shakhatreh, A.H. Sawalmeh, A. Al-Fuqaha, *et al.* Unmanned aerial vehicles (UAVs): a survey on civil applications and key research challenges. *IEEE Access*, 7, pp. 48572–48634, 2019.

[5] S. Hwang, J. Lee, H. Shin, S. Cho, and D.H. Shim. Aircraft detection using deep convolutional neural network for small unmanned aircraft systems. *AIAA Information Systems – AIAA Infotech @ Aerospace*, 2018.

[6] Comparing Deep Neural Networks and Traditional Vision Algorithms in Mobile Robotics by Andy Lee.

[7] O. Ronneberger, P. Fischer, and T. Brox. U-net: convolutional networks for biomedical image segmentation. In: N. Navab, J. Hornegger, W.M. Wells, and A.F. Frangi (eds.), *Medical Image Computing and Computer-Assisted Intervention – MICCAI 2015*, Cham: Springer International Publishing, 2015, pp. 234–241.

[8] D.C. Ciresan, L.M. Gambardella, A. Giusti, and J. Schmidhuber. Deep neural networks segment neuronal membranes in electron microscopy images. In: *NIPS*, 2012, pp. 2852–2860.

[9] A. Parkin. *Computing Colour Image Processing*. Cham: Springer, 2018.

[10] S.K. Ghosh. *Digital Image Processing*. Oxford: Alpha Science Intl Ltd, 2012.

[11] S.S. Haykin. *Neural Networks and Learning Machines,* 3rd ed. Upper Saddle River, NJ: Pearson Education, 2009.

[12] C.C. Aggarwal. *Neural Networks and Deep Learning*. New York, NY: Springer International Publishing Company, 2018.

[13] Airport Dataset, SenseFly Parrot Group.

[14] A.J. Pathak, M. Pandey, and S. Rautaray. Application of deep learning for object detection. *Procedia Computer Science*, 132, pp. 1706–1717, 2018. *International Conference on Computational Intelligence and Data Science.*

[15] U. Alganci, M. Soydas, and E. Sertel. Comparative research on deep learning approaches for airplane detection from very high-resolution satellite images. *Remote Sensing*, 12(3), pp. 458, 2020.

[16] Tzutalin. LabelImg Software. LabelImg: Graphical Image Annotation Tool. *MIT License*, 2015.

[17] M. Sandler, A. Howard, M. Zhu, A. Zhmoginov, and L. Chen. Mobilenetv2: inverted residuals and linear bottlenecks. In: *2018 IEEE/CVF Conference on Computer Vision and Pattern Recognition*, 2018, pp. 4510–4520.

[18] A.G. Howard, M. Zhu, B. Chen, *et al.* Mobilenets: efficient convolutional neural networks for mobile vision applications. *CoRR*, 2017, abs/1704.04861.

[19] W. Liu, D. Anguelov, D. Erhan, *et al.* Ssd: single shot multibox detector. In: B. Leibe, J. Matas, N. Sebe, and M. Welling (eds.), *Computer Vision – ECCV 2016*. Cham: Springer International Publishing, 2016, pp. 21–37.

[20] A. Bassiouny and M. El-Saban. Semantic segmentation as image representation for scene recognition. In: *2014 IEEE International Conference on Image Processing (ICIP)*, 2014, pp. 981–985.

[21] G. Chhor, C.B. Aramburu, and I. Bougdal-Lambert. Satellite image segmentation for building detection using U-Net, 2017. Online: http://cs229. stanford.edu/proj2017/final-reports/5243715.pdf.

[22] Chengwei. How to train an object detection model, 2019. Zhixuhao. Implementation of deep learning framework – unet, using keras.

[23] T.-Y. Lin, M. Maire, S. Belongie, *et al.* Microsoft coco: common objects in context. In: D. Fleet, T. Pajdla, B. Schiele, and T. Tuytelaars (eds.), *Computer Vision – ECCV 2014*. Cham: Springer International Publishing, 2014, pp. 740–755.

[24] D. Erhan, C. Szegedy, A. Toshev, and D. Anguelov. Scalable object detection using deep neural networks. In: *2014 IEEE Conference on Computer Vision and Pattern Recognition*, pp. 2155–2162, 2014.

[25] R. Padilla, W.L. Passos, T.L.B. Dias, S.L. Netto, and E.A.B. da Silva. A comparative analysis of object detection metrics with a companion open-source toolkit. *Electronics*, 10, p. 279, 2021.

[26] Y. Ho and S. Wookey. The real-world-weight cross-entropy loss function: modeling the costs of mislabeling. *IEEE Access*, 8, pp. 4806–4813, 2020, doi:10.1109/ACCESS.2019.2962617.

[27] A. Bharatha, K.H. Zou, S.K. Warfield, *et al.* Statistical validation of image segmentation quality based on a spatial overlap index, *Academic Radiology*, 11(2), pp. 178–189, 2004.

Part II

Rare event detection using Earth Observation data

Chapter 7

A transfer learning approach for hurricane damage assessment using satellite imagery

Jayesh Soni[1], Nagarajan Prabakar[1] and Himanshu Upadhyay[2]

Natural disasters such as hurricanes, wildfires, flooding, and landslides have become more frequent and severe due to climate change. Such disasters can damage building structures dramatically. Allocating resources and reconstructing buildings are typical disaster responses requiring fast and accurate assessment of affected areas. This process can be labour intensive. Traditional response strategies rely on building sensors detecting vibrations and in-field surveys, prohibitively expensive to deploy at large scales or overly time-consuming for processing accurate information. However, with the increased availability of satellite imagery, the task of understanding the structural damage to buildings can be performed in a remote and automated fashion using state-of-the-art computer vision approaches. This chapter addresses the problem of assessing the damage from hurricanes using satellite imagery. We provide an in-depth understanding of the convolution neural network (CNN), a neural network used for image datasets. Next, we describe various hyper-optimization techniques to optimize the training of the CNN network. Training the CNN network is often computationally expensive and time-consuming. Therefore, we present the transfer learning approach using the pre-trained model to solve the problem. We go over various transfer learning models such as AlexNet and VGGNet with their architectures. Next, we implement hurricane damage assessment by proposing a transfer learning-based framework on the open-source satellite image dataset of damages after a hurricane. This dataset is available on Kaggle. The dataset contains satellite images of the Greater Houston area after Hurricane Harvey in 2017. Each image has been labelled as either 'Flooded/Damaged' or 'Undamaged'. The proposed framework uses AlexNet to classify the images using advanced deep learning frameworks like TensorFlow. This approach can be extended to classify the satellite images of other natural calamities.

[1]Knight Foundation School of Computing and Information Sciences, Florida International University, USA
[2]Applied Research Center, Florida International University, USA

7.1   Introduction

Assessment of damage after a hurricane disaster is progressively more imperative for emergency administration. The existing method relies on disaster response crew members to drive into the affected area to analyse the damage using the windshield analysis approach. This method is expensive and time-intensive. Numerous experimental studies have been performed to support image analysis and further decrease the time for collecting data to speed up the assessment execution time. Learning-based algorithms are one of the noteworthy trends to distinguish whether a building is impaired or safe, after a hurricane landfall, on metrics such as accuracy, precision, and F1 score using satellite imagery [1]. Other research applied such learning algorithms to analyse panchromatic imagery and the time series of Landsat5/7 satellite imagery [2,3]. Chen *et al.* use synthetic aperture radar (SAR) data and extract the texture feature of the building to perform learning-based analysis. Deep learning techniques [4,5] have shown promising results within the field of damage assessment. Analysis of precipitation, a network of rivers [6], and usage of topological data to analyse the depth [7] were a few of the approaches used before deep learning-based CNN algorithm development. To detect impairment in the car [8,9], concrete structures [10–12], or detecting the change in the region after the impact of a hurricane [13], CNN is used extensively. Nevertheless, such methods require labelled datasets of good quality and quantity, which might be expensive in a few cases. Transfer Learning is the key to mitigating this issue to a reasonable extent. It is a method to develop and train the CNN model on a good dataset and use it for other related datasets. This improves training time and efficiency. The rest of the chapter is summarized as follows. Section 7.2 provides the literature review. Section 7.3 details the image processing technique and an in-depth overview of a deep learning-based CNN. Section 7.4 explains transfer learning with the various pre-trained model. Section 7.5 describes the practical use cases of the VGG-16 for damage classification. Finally, we conclude in Section 7.7.

7.2   Literature review

In this segment, we review advanced approaches in hurricane damage assessment. Mehrotra *et al.* [14] study the tsunami triggered by an earthquake in coastal areas of Japan to categorize pixels areas into bare land, vegetation, and water. They used earth elevation and SAR imagery data to perform this task. Nonetheless, such images are difficult for non-technical (e.g., emergency managers and first responders) to understand. Furthermore, optical sensors are installed in most artificial satellites than SAR sensors. Images of ground level [15] and satellite imagery [16,17] are the two distinct types of data that captures the damage to the building by the hurricane. Weber and Kané [18] utilize the images of the xBD database to predict the level of damage by developing and training the Mask R-CNN algorithm [19]. Hao *et al.* [20] used the xView2 dataset [21] containing a pair of satellite images to analyse the level of damage to the buildings. They developed a multi-class learning model. Cheng *et al.* [22] used an in-house dataset from Hurricane Dorian to train a stacked CNN. Hao and Wang [23] analyse the images of social networking platforms to detect the type of damage and level of severity by training five different machine learning algorithms. Recently, a transfer learning-based approach has been explored for utilizing the pre-trained model to assess hurricane damage. Many studies use aerial images [24–26] and apply transfer learning on a pre-trained CNN. This chapter proposes a transfer learning-based framework to assess hurricane damage using the open-source satellite image dataset.

7.3 Image processing techniques

There are two types of image analysis techniques as depicted in Figure 7.1.

The architecture of the can is roughly sketched as consisting of a bottom sensor layer, a middle network layer, and a top application layer. As one of the primary information-acquiring means at the bottom layer of the tags has found increasingly widespread applications in various business areas, with the expectation that the use of RFID tags will eventually replace the existing bar codes in all business areas.

7.3.1 Statistical-based algorithms

Open Source Computer Vision (OpenCV): It is one of the prevalent libraries for image processing library in the computer vision research area. It has varied functionality in numerous formats for transforming and shaping the input image into the desired output format. The image is stored in pixel format in the computer in Red, Blue, and Green (RGB) pixelated values. The computer only sees the array of numbers and not the actual image. There are three different unique pixel values in a three-dimensional data array for colour images. Height and width are the first two-dimensional value. The third dimension represents the percentage of RGB for each pixel. The pixel value in each channel ranges from 0 to 255, whereas for black and white images, the pixel value is in the range of 0–1.

7.3.2 Learning-based algorithms

Machine learning algorithms [27] learn the patterns from the dataset for prediction. There are three types of machine learning algorithms as defined below.

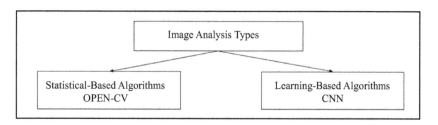

Figure 7.1 Image processing techniques

Supervised learning

Input data with a label is used for training purposes. Based on the target label value, such a learning algorithm is divided into classification and regression. Classification is when the output label is categorical, and the problem is a regression when the output label is a real value. Decision Trees, Random Forest, K Nearest Neighbour, Support vector machine, and linear regression are supervised learning algorithms.

Unsupervised learning

Input data without the target label is used for training. Such algorithms learn the hidden patterns from the unlabelled dataset. Clustering is one such type. It clusters the data into different groups. K-Means, Agglomerative, and Divisive are some of the unsupervised algorithms.

Semi-supervised learning

It contains a small part of the labelled dataset, with most data being unlabelled. The algorithms learn from the labelled data to predict unknown samples.

With the increase in computational power and rapid amount of data generation, deep learning algorithms are used heavily in almost every application area. One such deep learning-based algorithm is a CNN as shown in Figure 7.2. For image analysis such as object segmentation, object recognition and video processing, CNNs are widely used.

Let us, deep dive, into the working mechanism of CNN.

CNNs consist of four layers:

(i) *Convolutional layer:* This layer extracts the meaningful information or features from the image by applying a convolution operation when an input image is fed to this layer. Numerous such convolution operations extract the diverse distinct features from the image.

(ii) *Pooling layer:* The second layer after the convolutional layer. It takes the max or min or any statistical value to further reduce the dimension of the image. It is then passed to activation functions such as rectifier linear unit (ReLU) to have some nonlinearity.

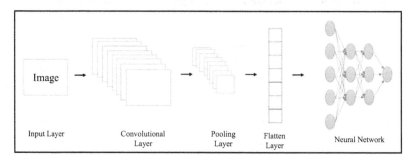

Figure 7.2   CNN

(iii) *Flatten layer:* This is the third layer in the architecture. The main objective of this layer is to flatten the input coming from the pooling layer.

(iv) *Neural network:* Finally, the last layer is a fully connected dense layer. The output of the flatten layer is processed by the neural network for classification, image segmentation, or object detection purposes.

Hyper-parameters of CNN

Every learning-based algorithm has hyper-parameters that need to be tuned to achieve optimal results. The values for the hyperparameters have to be set before training the model, and it is one of the active research areas. Let us discuss a few of the crucial hyperparameters for the CNN algorithm.

1. **Epochs**: Number of times the entire dataset is processed by the model during the training phase.

2. **Learning rate**: This parameter controls the rate of change in gradient descent, thus optimizing the weights. It can either be increased or decreased gradually or be kept fixed throughout the training time. It depends on popular optimizers such as Adaptive Delta, Adaptive momentum, RMSprop, and Stochastic Gradient Descent.

3. **Activation function**: They are used during the training of the neural network to apply nonlinearity. There are various activation functions, and the choice depends on the task that needs to be solved. ReLU is popularly used in CNN networks. Tangent hyperbolic and sigmoid are other activation functions that can be used.

4. **Batch size**: It controls the update of the weight during each epoch. It is usually in the range of 32–256 rows of the dataset, which can vary. It means the update of weights will occur after the processing of each batch of rows within the epoch.

5. **The number of hidden layers and units**: The total number of neurons in each hidden layer with the total number of hidden layers parameters is decided based on how well the model performs on the dataset. There is an underfitting issue with a small subset of it, whereas the model tends to be overfitted when such parameters have high values. Regularization techniques can be used to find the trade-offs.

6. **Dropout for regularization**: Dropout is used to prevent the overfitting of the model. If the dropout value is 0.3, 30% of the neurons in that layer will be deactivated during model training.

7. **Weight initialization**: A small set of random numbers can initialize the weights. Such random numbers should be uniformly distributed.

8. **Kernel size**: It is the filter's size used to extract features. Different kinds of features can be extracted from different sizes of the kernel.

9. **Stride**: It indicates how many pixel values the kernel should be moved while moving the kernel through the image in the convolution layer.

10. **Pooling**: It reduces the dimensionality of the features [28,29]; max-pooling will use the max value to reduce the feature, whereas min-pooling will use the min value to reduce the features.

Manual tuning of the hyper-parameters is a time-consuming process. Therefore, grid search and randomized search techniques can be employed to speed up the tuning process.

(A) Grid search

Grid search is an exhaustive search that trains the model on all the parameter combinations value specified during the model's training. It finds the best optimal parameter values for the particular data with the specified neural architecture, but on the other hand, it is computationally expensive. The computational time increases with the increase in the size of the neural network in terms of its hidden layer and the number of neurons in each hidden layer.

(B) Random search

Parameter selection is sampled randomly from a specific distribution, and thus, it is more computationally efficient than grid search. CNN has many hyper-parameters to tune. Also, it is time-consuming to find the optimal neural architecture. Therefore a technique called transfer learning has become very popular in recent years.

7.4 Transfer learning

Transfer learning is a method where a new problem can be solved using the pre-trained model on another problem domain, as depicted in Figure 7.3.

The significant advantage of this approach is reducing the neural network training time. The weights and biases of pre-trained models solve the new problem.

Advantages
1. *Meaningful features:* The first few layers of CNN learn to extract some basic features of the images for any other problem.
2. *Easy to access:* Such pre-trained models are accessible free of cost where we can download the weights through libraries providing application programming interface (APIs).
3. *Optimal results:* The models attained outstanding performance and are operational on the specific image recognition task for which they were developed.

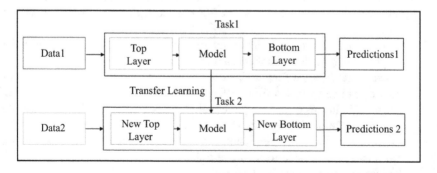

Figure 7.3   Transfer learning approach

Access to a large dataset of 1.2 million training images, 50k validation images, and 100k test images was provided to various teams in a competition for The ImageNet Large Scale Visual Recognition Challenge (ILSVRC) [30] to develop and train the model for classifying the images. There were a total of 1,000 categories. This open-source benchmark dataset is used to validate the models developed for transfer learning purposes.

Let us discuss various pre-trained models.

7.4.1 AlexNet

AlexNet [31] as shown in Figure 7.4 was developed in 2012 and was the winner of the ILSVRC competition, where it has decreased the error rate to half. It was the breakthrough towards the advancement of CNN networks. They utilize GPUs for the ReLU activation function. AlexNet architecture contains five convolutional layers and three fully connected layers with 60,000 parameters. Furthermore, the images were augmented by flipping, scaling, adding noise, etc.

It leads to the stride of length 4 in the pooling layer to reduce the error rate.

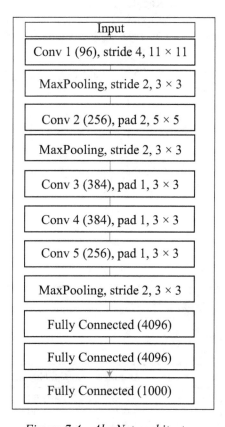

Figure 7.4   AlexNet architecture

7.5   Implementation

This section discusses the practical implementation of using AlexNet for damage classification. The following libraries are used for experimentation purposes.

Tensorflow: Tensorflow is an open-source library from google for deep learning and machine learning [32].
Keras: Keras is a wrapper on Tensorflow used by many societies worldwide. The code written in Keras is internally converted to TensorFlow for further execution. It has functional API and Sequential API [33].
Scikit-learn: Scikit-learn deals with a wide variety of learning-based algorithms (both supervised and unsupervised) [34].

Figure 7.5 shows the high-level framework for the implementation. It has three sections: Data Collection, Data Preprocessing, and Training Algorithm with final testing.

Stage 1: Data collection

This dataset is obtained from Kaggle. It contains satellite images of Hurricane Harvey in the Texas region. The training, validation, and test data are depicted in Table 7.1. Figures 7.6 and 7.7 show the sample images of No_Damage and Damage types, respectively.

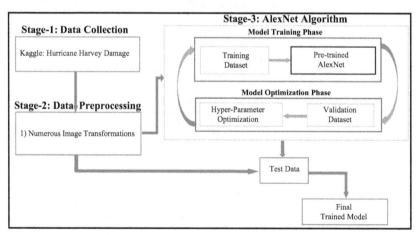

Figure 7.5   Proposed framework

Table 7.1   Dataset

Type	Damage	No_damage
Train	5,000	5,000
Validation	1,000	1,000
Test	1,000	1,000

 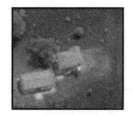 

Figure 7.6   Samples of no damage images

 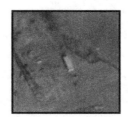 

Figure 7.7   Samples of damage images

Stage 2: Data pre-processing

The following pre-processing task using ImageDataGenerator from Keras is performed on the dataset:

1. Width shift range with a value of 0.3
2. Height shift range with a value of 0.3
3. Zoom range with a value of 0.2
4. Horizontal flip is set to true

Stage 3: Alexnet algorithm

3.1. Model Training Phase

We use the AlexNet CNN from the Keras library to train the model. To update the weight of the network, a mini-batch gradient descend optimizer is employed with a batch size of 64. BatchSize is a hyper-parameter that needs to be tuned while training the deep neural network. There are three variants of setting the batch size values. They are batch gradient descent, stochastic batch gradient descent, and mini-batch gradient descent. In batch gradient descent, we update the weights at the end of each epoch. For stochastic gradient descent, we update the weights after processing each data point. In mini-batch gradient descent, we update the weights at the end of every batch of data.

We used the mini-batch gradient descent algorithm and trained the model with batch sizes of {16, 32, 64, and 128}. We found out that batch size 64 gives the highest accuracy of 0.895.

Figure 7.8 shows the accuracy value at each epoch, whereas Figure 7.9 shows the loss at each epoch. Our trained model gives 0.89 accuracy with a loss of accuracy of 0.39 whereas VGGNet achieves an accuracy of only 0.75 [35].

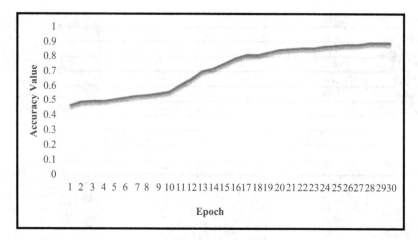

Figure 7.8   Accuracy with epoch

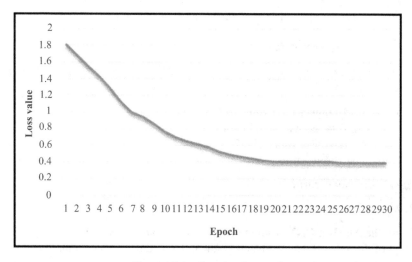

Figure 7.9   Loss with epoch

3.2. Model optimization phase

We used the validation dataset to perform the hyperparameters optimization of the model. The following hyper-parameters are tuned:

- Epochs: Number of times a model is trained on the entire dataset.
- Batch-size: Number of data points used to train the model before each weight update of the model.

Once we train the model with fine-tune optimization, we evaluate it with our test data and generate the final testing accuracy. This final trained model can be used for actual building damage assessment.

Table 7.2 Evaluated metrics

Accuracy	Precision	Recall	F1-score	ROC-AUC
0.895	0.894	0.852	0.867	0.869

For faster and optimal training, the ReLU activation function is used in all the convolution layers. Table 7.2 shows the evaluated metrics. Thus, we can conclude that a transfer learning algorithm can be applied to solve the problem where the dataset is limited and there is a limited computational resource. The model can further be deployed for real-time use cases. We performed all our experiments on Google Colaboratory, where we used Keras with TensorFlow at the backend to train the model and seaborn for visualization purposes.

7.6 Conclusion

Satellite images can prominently enable recovery and rescue efforts after the landfall of a hurricane. Numerous machine learning-based supervised classification is being applied to assess the damage to the building, and their use remains challenging. Since they require a good amount of labelled datasets, thus this approach is time-intensive. In this chapter, we provide an overview of CNN, a deep learning-based algorithm heavily used to solve the problem of image datasets. Every learning-based algorithm has many parameters that need to be tuned, and there is no precise formula to choose these values. Therefore, we presented a transfer learning-based approach where various pre-trained CNN-based algorithms can be applied directly to the new problem domain. We provide architectural details of several transfer learning-based algorithms and, finally, we presented a high-level implementation of building damage classification using an Alexnet pre-trained network.

This work can be extended by employing an ensemble approach by training multiple pre-trained algorithms to increase the model's accuracy. Additionally, we can utilize geological data to enhance the study of building damages.

References

[1] Q.D. Cao, and Y. Choe. Building damage annotation on post-hurricane satellite imagery based on convolutional neural networks. *Natural Hazards*, vol. 103, no. 3, pp. 3357–3376, 2020.
[2] C. Van der Sande, S. De Jong, and A. De Roo. A segmentation and classification approach of ikonos-2 imagery for land cover mapping to assist flood risk and flood damage assessment. *International Journal of Applied Earth Observation and Geoinformation*, vol. 4, no. 3, pp. 217–229, 2003.
[3] S. Skakun, N. Kussul, A. Shelestov, and O. Kussul. Flood hazard and flood risk assessment using a time series of satellite images: a case study in Namibia. *Risk Analysis*, vol. 34, no. 8, pp. 1521–1537, 2014.

[4]   J. Soni, N. Prabakar, and H. Upadhyay. Behavioral analysis of system call sequences using LSTM seq-seq, cosine similarity and jaccard similarity for real-time anomaly detection. In *2019 International Conference on Computational Science and Computational Intelligence (CSCI)*. IEEE, 2019, pp. 214–219.

[5]   J. Soni, N. Prabakar, and H. Upadhyay. Comparative analysis of LSTM sequence–sequence and auto encoder for real-time anomaly detection using system call sequences. *International Journal of Innovative Research in Computer and Communication Engineering*, vol. 7, no. 12, pp. 4225–4230, 2019.

[6]   H. Apel, A.H. Thieken, B. Merz, and G. Bloschl. Flood risk assessment and associated uncertainty. *Natural Hazards and Earth System Science*, vol. 4, no. 2, pp. 295–308, 2004. Available: https://hal.archives-ouvertes.fr/hal-00299118.

[7]   J. W. Hall, R. Dawson, P. Sayers, C. Rosu, J. Chatterton, and R. Deakin. A methodology for national-scale flood risk assessment. In *Proceedings of the Institution of Civil Engineers-Water and Maritime Engineering*, vol. 156, no. 3. Thomas Telford Ltd, 2003, pp. 235–247.

[8]   K. Patil, M. Kulkarni, A. Sriraman, and S. Karande. Deep learning based car damage classification. In *2017 16th IEEE International Conference on Machine Learning and Applications (ICMLA)*. IEEE, 2017, pp. 50–54.

[9]   D. Zhang, Y. Zhang, Q. Li, T. Plummer, and D. Wang. Crowdlearn: a crowd-ai hybrid system for deep learning-based damage assessment applications. In *2019 IEEE 39th International Conference on Distributed Computing Systems (ICDCS)*. IEEE, 2019, pp. 1221–1232.

[10] Y.J. Cha, W. Choi, and O. Büyüköztürk. Deep learning-based crack damage detection using convolutional neural networks. *Computer-Aided Civil and Infrastructure Engineering*, vol. 32, no. 5, pp. 361–378, 2017.

[11] Y.J. Cha, W. Choi, G. Suh, S. Mahmoudkhani, and O. Büyüköztürk. Autonomous structural visual inspection using region-based deep learning for detecting multiple damage types. *Computer-Aided Civil and Infrastructure Engineering*, vol. 33, no. 9, pp. 731–747, 2018.

[12] H.-w. Huang, Q.-t. Li, and D.-m. Zhang. Deep learning-based image recognition for crack and leakage defects of metro shield tunnel. *Tunnelling and Underground Space Technology*, vol. 77, pp. 166–176, 2018.

[13] J. Doshi, S. Basu, and G. Pang. From satellite imagery to disaster insights, 2018. arXiv preprint arXiv:1812.07033.

[14] K.K. Mehrotra, M.J. Singh, Nigam, and K. Pal. Detection of tsunami-induced changes using generalized improved fuzzy radial basis function neural network. *Natural Hazards*, vol. 77, no. 1, pp. 367–381, 2015.

[15] D.B. Roueche, F.T. Lombardo, R. Krupar, and D.J. Smith. *Collection of Perishable Data on Wind-and Surge-Induced Residential Building Damage During Hurricane Harvey (TX)*. Austin, TX: DesignSafe-CI, 2018.

[16] D. Lam, R. Kuzma, K. McGee, *et al.* xview: Objects in context in overhead imagery, 2018. arXiv:1802.07856.

[17] R. Gupta, R. Hosfelt, S. Sajeev, *et al.* xbd: a dataset for assessing building damage from satellite imagery, 2019. arXiv:1911.09296.

[18] E. Weber and H. Kané. Building disaster damage assessment in satellite imagery with multi-temporal fusion, 2020. *arXiv:2004.05525.*

[19] K. He, G. Gkioxari, P. Dollár, and R. Girshick. Mask r-cnn. In *Proceedings of the IEEE International Conference on Computer Vision*, Venice, Italy, 27–29 October 2017, pp. 2961–2969.

[20] H. Hao, S. Baireddy, E.R. Bartusiak, *et al.* An attention-based system for damage assessment using satellite imagery, 2020. arXiv:2004.06643.

[21] R. Gupta, B. Goodman, N. Patel, *et al.* Creating xBD: a dataset for assessing building damage from satellite imagery. In *Proceedings of the IEEE/CVF Conference on Computer Vision and Pattern Recognition Workshops*, Long Beach, CA, USA, 16–17 June 2019, pp. 10–17.

[22] C.S. Cheng, A.H. Behzadan, and A. Noshadravan. Deep learning for post-hurricane aerial damage assessment of buildings. *Computer-Aided Civil and Infrastructure Engineering,*, vol. 36, pp. 695–710, 2021.

[23] H. Hao and Y. Wang. Hurricane damage assessment with multi-, crowd-sourced image data: a case study of Hurricane Irma in the city of Miami. In *Proceedings of the 17th International Conference on Information System for Crisis Response and Management (ISCRAM)*, Valencia, Spain, 19–22 May 2019.

[24] Y. Li, W. Hu, H. Dong, and X. Zhang. Building damage detection from post-event aerial imagery using single shot multibox detector. *Applied Science*, vol. 9, p. 1128, 2019.

[25] M. Presa-Reyes and S.C. Chen. Assessing building damage by learning the deep feature correspondence of before and after aerial images. In *Proceedings of the 2020 IEEE Conference on Multimedia Information Processing and Retrieval (MIPR)*, Shenzhen, China, 6–8 August 2020, pp. 43–48.

[26] Y. Pi, N.D. Nath, and A.H. Behzadan. Convolutional neural networks for object detection in aerial imagery for disaster response and recovery. *Advanced Engineering Informatics*, vol. 43, p. 101009, 2020.

[27] J. Soni and N. Prabakar. Effective machine learning approach to detect groups of fake reviewers. In *Proceedings of the 14th International Conference on Data Science (ICDATA'18)*, Las Vegas, NV, 2018, pp. 3–9.

[28] J. Soni, N. Prabakar, and H. Upadhyay. Visualizing high-dimensional data using t-distributed stochastic neighbor embedding algorithm. In *Principles of Data Science*. Cham: Springer, 2020, pp. 189–206.

[29] J. Soni, N. Prabakar, and H. Upadhyay. Feature extraction through deep-walk on weighted graph. In *Proceedings of the 15th International Conference on Data Science (ICDATA'19)*, Las Vegas, NV, 2019.

[30] O. Russakovsky, J. Deng, H. Su, *et al.* Imagenet large scale visual recognition challenge. *International Journal of Computer Vision,* vol. 115, no. 3, pp. 211–252, 2015.

[31] Md.Z. Alom, T.M. Taha, C. Yakopcic, *et al.* The history began from alexnet: a comprehensive survey on deep learning approaches, 2018. preprint arXiv:1803.01164.

[32] J.V. Dillon, I. Langmore, D. Tran, *et al.* Tensorflow distributions, 2017. preprint arXiv:1711.10604.

[33] N. Ketkar. Introduction to keras. In *Deep Learning with Python*. Berkeley, CA: Apress, 2017, pp. 97–111.

[34] F. Pedregosa, G. Varoquaux, A. Gramfort, *et al.* Scikit-learn: machine learning in Python. *Journal of Machine Learning Research,* vol. 12, pp. 2825–2830, 2011.

[35] S. Kaur, S. Gupta, S. Singh, *et al.* Transfer learning-based automatic hurricane damage detection using satellite images. *Electronics,*; vol. 11, no. 9, p. 1448, 2022.

Chapter 8

Wildfires, volcanoes and climate change monitoring from satellite images using deep neural networks

Yash Kumar Shrivastava[1] and Kavita Jhajharia[1]

Climate change refers to the long-term change in weather patterns and temperatures, driven by multiple factors. Since 1880, human activities have been the primary accelerators of climate change. Climate change has broad, far-reaching impacts on the environment. Sometimes these impacts result in catastrophic disasters, causing huge economic (infrastructure and property loss, poverty, etc.) and social (human lives, animal lives, diseases, malnutrition, displacement of families, mental health issues, job losses, food, and water shortage, etc.) losses. These impacts include Wildfires, Volcanic Eruptions, Landslides, etc. In 2020, in the United States alone, there were 58,950 wildfires, burning about 10.12 million acres of land. In India, in 2019 alone, landslides were responsible for the loss of 264 lives. The current global warming situation only accelerates these disasters even more.

Monitoring the impacts of climate change will be crucial in tackling it, and in reducing the consequences and losses caused due to disasters. Active and efficient monitoring of disasters can help in early warning, which can ensure a faster and more effective response, saving countless lives and reducing losses. Satellite imagery plays a crucial role in monitoring climate change and observing its impacts around the world. Wildfires, landslides, volcanoes, etc., can all be identified and monitored with satellite imagery. Satellite imagery has historically been used to monitor the long-term effects of climate change, like changes in the ice cap, forest cover, etc. Satellite images of volcanic eruptions, wildfires, landslides, etc. help in mapping the data resulting in more efficient rescue operations.

Today, methods like deep learning (DL) and machine learning (ML) can play a crucial role in improving the efficiency of monitoring climate change and natural disasters. ML is a branch of Artificial Intelligence (AI) and computer science that involves the study of computer algorithms that can improve themselves through experience, observations, and the use of data. DL is a part of a broader family of ML methods that use multiple layers to progressively extract higher-level features

[1]Manipal University Jaipur, India

from raw input. Applying DL and convolutional neural network (CNN) techniques to satellite images can help in automating the disaster identification process, thereby making the process faster and much more efficient. Image segmentation can be of great help in monitoring environmental changes and natural disasters. Image segmentation is a process that involves the division of a digital image into various parts. These parts are called 'Image segments'. This process helps in reducing the complexity of the image and it also makes further processing and analysis of the image much more simplified. Implementation of image segmentation techniques on very high-resolution (VHR) satellite images can substantially help in monitoring climate change and natural disasters like wildfires, landslides, volcanic eruptions, etc.

Wildfires are spontaneous destructive fires that spread quickly over forests, woodland, and other combustible vegetation. CNNs can be used to help automate the detection process of wildfires using VHR satellite imagery after training the model with data. Image segmentation can be performed over satellite images for automated detection of wildfires based on training data. The model can be trained to divide the image into segments of fire and non-fire, burnt area and unburnt area. Data augmentation techniques can also be used to enlarge the training data set. CNNs like U-Net, Inception-v3, etc. have been used in the past and be very effective in classifying images. Synthetic aperture radar (SAR) can also play a critical role in wildfire detection and monitoring, since it can penetrate through clouds and smoke, and is also capable of imaging day and night. A CNNs-based DL framework can be used to identify the burnt area and differentiate it from the unburnt area.

Wildfires can also trigger many other disastrous hazards, one such particular is the occurrence of landslides in the burnt area. The burnt plots of land area are highly susceptible to debris flows. Satellite imagery has made it much easier to identify the areas affected by wildfires, and it has also enabled efficient assessment of burnt areas after the wildfire. Assessment of post-wildfire burnt areas is a very challenging task and poses a lot of risks. Satellite imagery makes this much easier and risk-free. Manual monitoring of landslides too is a very labour-intensive and time-consuming task. This problem can also be dealt with through DL methods, by automating the task of landslide classification through satellite imagery for more robust classification results. CNN's like U-Net can be very useful for classification purposes and landslide mapping can be done using VHR satellite images. These DL techniques can automate the task of landslide detection, hence reducing the time and human effort.

Volcano deformation is another phenomenon that can be monitored using DL techniques. Satellite data can help in large-scale global monitoring and identification of volcanic activity and provide the first indications of volcanic eruptions, instability, etc. Volcanic deformation is a very good indicator of volcanic eruptions as it usually happens before a volcanic eruption. Here, interferometric SAR (InSAR) can be very useful in monitoring volcanic activity, as it can help in plotting surface deformation. The images generated by InSAR are known as interferograms and they contain contributions from both deformation of volcanoes and radar paths through the atmosphere. ML models can identify useful signals from a large collection of InSAR data. Since the number of deforming volcanoes is very low, implementing ML-based monitoring solutions becomes even more difficult. CNNs like AlexNet can be very useful in automating this task. AlexNet can be trained using synthetic interferograms for this task. The synthetic interferograms contain deformation patterns based on multiple probability selection, effects caused by stratified atmospheres derived from weather models, and other atmospheric effects caused by irregularities in air motion. Such efficient DL models can help identify volcanic activity faster and give warnings before volcanic unrest.

8.1 Introduction

Climate change has increasingly become a cause of concern among experts and common citizens around the globe alike. But it must be noted that climate change is not a new concern and has been discussed for a long time. The earliest concerns were raised in 1938 by a British engineer who went by the name 'Guy Callendar' [1]. He compiled the records of 147 weather stations located around the globe and used the data to show that the temperatures had risen over the preceding century. He also noted that the concentration of carbon dioxide had increased significantly over the same period. He suggested that this increment in CO_2 levels was the primary reason for the rise in the temperatures. This theory was widely known as the 'Callendar Effect'. It was largely dismissed by experts at that time [2].

Today, several governments around the world have started taking stringent measures to keep a check on climate change and reduce its impact as much as possible, and as soon as possible. Several countries have set respective goals to reach the so-called 'Net-Zero Emissions' within a span of a few years. Countries around the globe have come together on different occasions and summits to set different goals for curbing climate change and reducing its impacts effectively and quickly. One such summit is the COP26 [3], hosted in March 2021 [4], in Glasgow, United Kingdom. COP – an abbreviation for 'Conference of Parties' – is the latest yearly climate change conference held by the United Nations. It was attended by all the countries that had signed the United Nations Framework Convention on Climate Change (UNFCCC). This framework came into effect in 1994. COP26 was the 26th summit of its kind. This summit saw many countries pledging the attainment of 'Net-Zero Emissions' within a fixed period. In simple terms, 'Net-Zero' emissions are achieved when the amount of carbon present in the atmosphere is equal to the amount of carbon that is being removed from the atmosphere. And 'Net-Positive' emissions are achieved when the amount of carbon present in the atmosphere is less than the amount of carbon that is being taken away from the atmosphere. Most countries pledged to attain Net Zero emissions before 2050. Some examples are Canada (2050), Germany (2045), and Nepal (2045), some countries have set longer deadlines, like India (2070) and China (2060). Some countries have already achieved Net Zero emissions (Mauritius, Bhutan, etc.). Some countries, including Japan, Canada, and the EU countries, have legally binding Net-Zero commitments. According to NASA's data, carbon-di-oxide levels have steeply risen over the last 100 years [5].

The temperature of the western coast of Antarctica has risen significantly over the past five decades at an alarming rate. The temperatures of the Upper Ocean to the west of Antarctica have risen by 1 degree centigrade since the year 1955. This warming of the Antarctic peninsula has caused a very serious impact on the Antarctic fauna and the living environment there. The overall penguin settlement distribution has changed significantly over the period owing to the rise in temperatures [6]. It is predicted that the oceanic habitat of the Antarctic Krill [7], a finger-sized aquatic organism that plays a critical role in the oceanic food web, may move south in the future, and this shift can also be associated with climate change. The habitat is also said to further deteriorate during the periods of summer and autumn. Greenland lost about 280 gigatons of Ice caps every year from the year 2002–2021, and this can be associated with climate change. This ice melting has resulted in an increase in global sea level by 8 mm per year during the same period. But the impact of climate change is not just seen in remote places like Antarctica and Greenland. The effects are very much felt in every corner of the world, just in slightly different forms. Disasters that are caused due to climate change, also known as 'climate-induced disasters (CIDs)', have increasingly become a cause of concern for nations around the globe. Such disasters are becoming more and more frequent and are occurring at a larger scale than ever before, claiming more lives and causing increasingly higher amounts of property damage.

According to the 2020 Global Risk Report [8], released by the World Economic Forum, climate-related risks claimed three out of the top five risks from 2018 to 2020, based on the likelihood and impact of the respective risks. Extreme weather events claimed the top spot in the same report. These CIDs range from wildfires to volcanic activity and everything in between [9]. Wildfires are spontaneous destructive fires that spread quickly over forests, woodland, and other combustible vegetation. In recent years, wildfires have become very frequent in occurrence. Several large wildfires have occurred in the last 4–5 years. The wildfires in Australia that occurred from 2019 to 2020 alone are estimated to have burnt around 24.3–33.8 million hectares of land. This ended up burning down an estimated 5,900 buildings, which included around 2,700 homes and ended up claiming 34 lives. An estimated 3 billion vertebrates living in the region were affected by these wildfires. These wildfires have cost massive economic loss to Australia. Around 103 billion dollars were lost on the property and economic losses. The Amazon Rainforest Wildfires that occurred around the same time period as the Australian bushfires have also been equally devastating. Amazon rainforest experienced more than 29,000 forest fires, causing significant economic and property losses. The Uttarakhand wildfires of 2016 are another such example. Wildfires in Uttarakhand have burnt around 110,000 acres of forest land since 2000. About 8,000 acres were lost in the 2016 forest fires alone, claiming at least seven lives. These incidents could have been predicted, and many human lives could have been saved, had there been enough systems in place to prevent and predict such events. Monitoring climate change and its impacts is the first step in the prevention of such CIDs. Monitoring wildfires has historically been a complex challenge. But with the latest advancements in technology, monitoring wildfires has become much easier.

The availability of data and resources like satellite imagery can be of immense help in monitoring and predicting the occurrence and the spread of wildfires [10], which in turn can be of great use to the authorities carrying out rescue operations, thus making the rescue process much smoother and easier. Modern DL methods can be immensely useful in analysing the data from satellite images and monitoring/predicting the occurrence or spread of wildfires. Today, methods like DL and ML can play a crucial role in improving the efficiency of monitoring climate change and natural disasters. ML is a branch of -AI and computer science, that involves the study of computer algorithms that can improve themselves through experience, observations, and the use of data. DL is a part of a broader family of ML methods that use multiple layers to progressively extract higher-level features from raw input [11]. Applying DL and CNN techniques to satellite images can help in automating the disaster identification process, thereby making the process faster and much more efficient. Image segmentation can be a great way of monitoring environmental changes and natural disasters. Image segmentation is a process that involves the division of a digital image into various parts. These parts are called 'Image segments'. This process helps in reducing the complexity of the image and it also makes further processing and analysis of the image much more simplified. Implementation of image segmentation techniques on VHR satellite images can substantially help in monitoring climate change and natural disasters like wildfires, landslides, volcanic eruptions, etc. CNNs can be used for the automatic detection of wildfires using VHR satellite imagery after training the model with data. Image segmentation can be performed over satellite images for automated detection of wildfires based on training data. The model can be trained to divide the image into segments of fire and non-fire, burnt area, and unburnt area. Data augmentation techniques can also be used to enlarge the training data set [12]. SAR can also play a critical role in wildfire detection and monitoring since it can penetrate through clouds and smoke, and is capable of imaging day and night. A CNNs-based DL framework can be used to identify the burnt area and differentiate it from the unburnt area. Wildfires are only one of the many CIDs that occur in different parts of the world. One such CID is landslide. Landslides occur when a mass of rock, earth, or debris slides down a slope of the land. Landslides cause catastrophic economic loss and claim many lives every year.

According to the World Health Organization (WHO), between the years 1998 and 2017, approximately 4.8 million people were affected by landslides. Landslides claimed around 18,000 lives during that same year. It has become extremely important to monitor landslides. Landslides can be triggered by wildfires [13]. The burnt plots of land area are highly susceptible to debris flows. Thus, preventing wildfires from occurring can hugely reduce the number of landslides occurring every year. Satellite imagery has made it much easier to identify the areas affected by wildfires, and it has also enabled efficient assessment of burnt areas after the wildfire. Assessment of post-wildfire burnt areas is a very challenging task and poses a lot of risks. Satellite imagery makes this much easier and risk-free. CNN can be trained to classify the parts of the land into segments of burnt and unburnt areas, thus making it far easier to predict the occurrence of landslides, and thereby making rescue and evacuation operations much faster. Since manual monitoring of landslides is very challenging and labour intensive which poses a lot of risk to the personnel, such DL methods can be greatly useful in making the task of monitoring much easier. Assessment of post-wildfire burnt areas is a very challenging task and poses a lot of risks. Satellite imagery makes this much easier and risk-free and this problem can also be dealt with through DL methods, by automating the task of landslide classification through satellite imagery for more robust classification results. CNN like U-Net can be very useful for classification purposes and landslide mapping can be done using VHR satellite images. These DL techniques can automate the task of Landslide detection, hence reducing the time and human effort. Volcano deformation is another phenomenon caused by climate change. Volcano deformation is when the surface of a volcano and the land around a change shape. This is a very important indicator of volcanic eruptions [14].

Volcanic eruptions are extremely disastrous and cause a lot of economic and human resource loss. Monitoring volcano deformation is extremely important as it enables researchers and scientists to understand the situation of the volcano and estimate what is happening within it. This data can be very helpful in determining the possibility of a volcanic eruption. This can be very helpful in the case of an eruption as the people living around such an area can be informed in advance and hence the rescue and evacuation operations can be carried out in a much smoother and more efficient manner, hence, it will lead to many more lives being saved [15]. Modern DL methods can be very helpful in monitoring such volcanic activity with the help of satellite imagery and technologies like InSAR [16]. The images generated by InSAR are known as interferograms and they contain a contribution from both deformation of volcanoes and radar path through the atmosphere. DNNs like AlexNet can be trained to monitor InSAR data and warn about the occurrence of such deformations, so that action can be taken in time by the rescue authorities. AlexNet can be trained using synthetic interferograms. The synthetic interferograms contain deformation patterns based on a multiple probability selection, effects caused by stratified atmosphere derived from weather models, and other atmospheric effects caused by irregularities in air motion. Thus, an efficient DL model can identify volcanic activity and give warnings before volcanic unrest. Often, volcanic eruptions may also lead to wildfires, and wildfires can also lead to landslides. Hence, having monitoring systems in place can be crucial and can hugely reduce the impact caused by these CIDs, saving countless lives [17].

Cyclones are yet another major cause of concern around the globe. Cyclones cause a lot of economic damage and take many lives every year. In the last 50 years, around 14,942 calamities and disasters have been attributed to tropical cyclones. An estimated 779,324 people have lost their lives due to cyclone-related disasters. This number averages out to be around 43 deaths every day. These cyclones have caused 1.4 trillion dollars in economic losses, which is an average of 78 million dollars in losses every single day for those 50 years. Hence, monitoring cyclones and coastal water bodies is crucial. CNN-based models can be trained to identify and predict the possibility of a cyclone approaching the coast and can also predict its path and speed. This can be crucial in evacuation and rescue missions. Any time saved in such missions can save many more lives. Thus, the existence of such monitoring systems can be of great benefit. It will make the task of evacuation authorities a lot easier and will make the whole evacuation process a lot faster and smoother. If people living in coastal areas are alerted about the possibility of a cyclone hitting the area, the evacuation can start much before the actual calamity is about to occur, thus giving a lot more time for people to move themselves as well as their belongings to safer places. Monitoring cyclones has been attempted with UAV, but this method is unsustainable and impractical for a variety of reasons. First, while the cost of UAVs themselves is relatively low, their imaging elevation height is very low, and they need to get dangerously close to the cyclone. This may lead to accidents and the loss of drones. Second, since the drones need to be so close to the cyclone, they do not stay stabilized, and the film comes out to be very shaky and at times unusable. Another issue with drones is their flight time, satellites have a practically unlimited flying time, compared to the minutes of flight time drones have. This makes them very impractical to use. And drones are unusable for 24/7 monitoring of water bodies. Also, they have a very limited range, and hence, they are unviable for this purpose. Hence, satellites prove to be the best medium for monitoring water bodies and cyclones [18].

CNN-based models can be very useful in extracting deep features from live image feeds and can be used to predict the intensity of the cyclone and its path. A once crucial way of tracking a cyclone is by detecting its 'eye'. The eye of a cyclone is the area of the cyclone where the wind flow is extremely slow and is generally very calm. This 'eye' usually lies in the centre of the cyclone, and I am usually 30–35 km in diameter. The eye is surrounded by the 'eye wall' of the cyclone. This is the region where the severity of the cyclone is the highest and very fast winds blow in this region. DL models can be trained with images to identify these features in the images and estimate their severity, speed and path, based on data from prior cyclones.

8.2   Background and related work

While the issue of climate change is increasingly being talked about on a global stage, there has been quite a lot of work done already on this topic. Previous works have shown the application of DL methods to monitor climate change and its different impacts, and other climate-induced calamities. In this study, we have discussed the application of DL neural networks (DNNs) such as AlexNet and U-Net on satellite imagery to monitor climate change and CIDs. Previous studies have proven the efficiency of these neural networks, as well as some other DNNs in monitoring different aspects of climate change.

Previous research, titled 'Wildfire Segmentation on Satellite Images using Deep Learning', conducted by Vladimir Khryashchev and Roman Larionov of the P.G. Demidov Yaroslavl State University in Russia [12], discusses the applications of DL methods in wildfire segmentation using satellite images. It demonstrates the application of a CNN model called U-Net along with ResNet34 being used as an encoder. They have demonstrated how U-Net, which was initially introduced as a DNN for biomedical image segmentation, has applications in wildfire segmentation. They used the Planet dataset and the Resurs dataset for the Resurs-P satellite. They developed a U-Net-like model – U-ResNet34 and were able to achieve excellent results.

Another paper titled 'Deep Learning Based Forest Fire Classification and Detection in Satellite Images' by R. Shanmuga Priya and K. Vani of Anna University [19] discusses forest fire classification using CNNs. They have used inception-v3 and trained the model using satellite images to classify between fire and non-fire zones and burnt and un-burnt patches of land. They have used a transfer learning-based approach. They have also focused on smoke detection [20], as it is one of the primary indicators of a possible forest fire. Their fire detection model, based on inception-v3, was able to achieve 98% accuracy.

A paper titled 'A deep learning approach to detecting volcano deformation from satellite imagery using synthetic datasets', published by N. Anantrasirichaia, J. Biggs, F. Albino, and D. Bull of the University of Bristol [17] discusses applications of DL in monitoring volcanic deformation. They discuss how volcano deformation is a great indicator of future volcanic eruptions and how it is crucial to monitor this phenomenon. While working on this, they encountered a problem – as the no of such deforming volcanoes is relatively less, training traditional models becomes much more difficult. They tackled this issue by using synthetic inferograms to train the AlexNet CNN. Using synthetic inferograms to train AlexNet turned out to provide more accurate results, than using real inferograms. They ended up retaining the CNN with synthetic inferograms and some select real ones. The final PPV that they achieved turned out to be 82%.

8.3   Modern DL methods

DL is a type of ML, where multiple layers of processing are used to extract increasingly higher-level features from the given data. It is based on artificial neural networks (ANNs). Applying DL and CNN techniques to satellite images can help in automating the disaster identification process, thereby making the process faster and much more efficient. Image segmentation is an efficient and useful way of monitoring environmental changes and natural disasters, and changes on the land. Image segmentation is a process that involves the division of a digital image into various pieces. These pieces are called 'Image segments'. This process helps in reducing the complexity of the image and it also makes further processing and analysis of the image much more simplified and easier. Implementation of image segmentation techniques on VHR satellite images can substantially help in monitoring climate change and natural disasters like wildfires, landslides, volcanic eruptions, etc. This chapter deals with two different neural networks: U-Net and AlexNet. These neural networks are used to monitor climate change and its different impacts and CIDs.

8.3.1   U-Net

U-Net is a new neural network and was first used in 2015 for biomedical image segmentation. U-Net's primary use case is image segmentation. U-Net takes in an image as an input and it outputs a label. U-Net does its classification on every pixel, and hence the size of the input and out in U-Net is the same in every case. This property of U-Net is very useful in bio-medical image segmentation and anomaly detection and it enables both detection and localization of the anomaly. U-Net has many use cases in climate change monitoring, and especially the availability of satellite imagery makes the case even stronger for U-Net applications. UNet is a u-shaped neural network, and it contains an encoder and a decoder. These are the two main components of U-Net. This is a very useful neural network as it can achieve great accuracy in results with very little training data. U-Net finds its use in wildfire segmentation. A model can be trained to identify burnt and un-burnt land plots and hence makes the task of authorities carrying out rescue operations much easier.

8.3.2   AlexNet

AlexNet is a CNN model that solves a problem faced by many traditional neural networks – they are difficult to be applied to high-resolution images. Since the monitoring of climate change requires the usage of VHR satellite images, it makes perfect sense to apply AlexNet in this case. AlexNet is an incredibly powerful CNN model that can achieve high accuracies on very challenging datasets. The AlexNet architecture consists of eight convolutional layers. Five of these layers are convolutional layers while three are fully connected layers. One of the many unique AlexNet features that are not present in traditional CNNs is the ability to work with multiple graphical processing units (GPUs). Half of the model's neurons can be put on 1 GPU and half on another. This allows much bigger models to be trained and cuts down on the time required for training. AlexNet can be very useful in monitoring volcanic deformation using VHR satellite imagery, primarily because AlexNet's application on high-resolution images is much easier and more convenient than traditional CNNs [21].

8.3.3   Inception-v3

Inception-v3 in a CNN model that is used for image segmentation. It has achieved an accuracy rate of 78.1% on the ImageNet database. The Inception-v3 is also less power/resource-consuming in comparison with the older versions. It has a total of 42 layers and the error rate is much lower than that of its earlier iterations/versions. Inception-v3 is used in the classification of VHR satellite images using image segmentation [22].

8.3.4   Other neural networks

SqueezeNet [23], originally released in 2016, is a deep neural network (DNN) with 18 layers and can be very useful in detecting fire in satellite images. Another useful

DNN is ResNet. ResNet is an abbreviation for 'residual network'. It is called so because it uses a technique called 'residue mapping' to combat the plateauing and a gradual degradation in model accuracy caused during training. ResNet [24] is useful in smoke detection during wildfires.

8.4   Benefits of using this approach

Applications of DL methods in climate change monitoring have immense advantages and benefits over conventional methods. A few benefits have been discussed below. DL methods provide far more accurate results than conventional methods. Many models discussed in this study have been able to achieve extremely high accuracy rates. An Inception-v3-based model, used for the segmentation of burnt and un-burnt land after a wildfire, has achieved an accuracy rate of 98%. This makes these modes extremely reliable and much better to use in comparison with conventional models. DL methods are far less time-consuming. They are capable of instantly identifying a change or a disturbance and instantly alerting the respective authorities. This enables the authorities to take action as soon as possible, which ends up saving a lot of lives in many cases. It makes rescue and evacuation operations much smoother and gives the personnel enough time to evacuate as many people as they possibly can.

DL models can monitor 24/7 and they rarely need any repairs/changes. This enables a 24/7 monitoring and alerting system to exist that can detect a calamity at any time of the day and can instantly alert the respective authorities. This is much better than conventional models as people monitoring may need rest and cannot work 24/7. These setups require very little human interaction. This hugely reduces labour costs as very limited human input is required on rare occasions. These algorithms are largely automated. And this is precisely why the algorithms can work so efficiently and without any breaks. This ends up saving a lot of human capital, and monetary expenses. These models keep improving their detection skills as time goes by; hence, the detection system becomes much more reliable and faster as it ages. This is very different from conventional set-ups where humans face a reduction in ability and efficiency as time goes by.

These DL models can predict ground changes and the possibility of the occurrence of natural calamities with extremely high accuracy. This perhaps is the biggest point of difference between the conventional approach and the DL approach, and, in this case, the DL approach works out to me as a much better option. Predictions can not only help rescue and evacuation authorities prepare for such events in advance but they can also be used to alert citizens in advance and immensely reduce the effort of authorities. The overall input and set-up costs for implementing these models are much lower than, say, the cost of setting up conventional weather stations for example. Such stations need to hire personnel and pay them, buy systems, and office furniture and rent out large office spaces. While our approach can be implemented in relatively smaller rooms with much fewer personnel and much fewer systems.

8.5   Long-term climate change monitoring using DL methods

Climate change refers to the long-term change in weather patterns and temperatures, driven by multiple factors. Since 1880, human activities have been the primary accelerators of climate change. Climate change has broad, far-reaching impacts on the environment. Sometimes these impacts result in catastrophic disasters, causing huge economic (infrastructure and property loss, poverty, etc.) and social (human lives, animal lives, diseases, malnutrition, displacement of families, mental health issues, job losses, food, and water shortage, etc.) losses. These impacts include wildfires, volcanic eruptions, landslides, etc. In 2020, in the United States alone, there were 58,950 wildfires, burning about 10.12 million acres of land. In India, in 2019 alone, Landslides were responsible for the loss of 264 lives. The current global warming situation only accelerates these disasters even more. CNN-based models can be extremely useful in monitoring the climatic changes around the world and this can be helpful in many ways. Many countries have made commitments at several climate conventions, to curb climate change. One of the many ways these countries can use to keep track of the progress they have made in curbing climate change can be with the use of satellite images.

Moreover, DL models can be applied to satellite images to monitor the progress. One of the many use cases of DL in this area would be in the analysis of air quality using satellite images. Models can be trained with images of days when the air quality index of the specific cities was low, and images when the air quality of the same areas of the same cities was high. This can be extremely useful in monitoring the air quality index of remote places where air quality measurement is not possible. This can be crucial in a country's progress towards its respective climate goal. Another use case of such models can be to look at concerning signs of climate change and alert the respective authorities to take action or change policies. Several CIDs that we have discussed previously avoided had some action taken before. Climate change is responsible for many such calamities and looking for concerning and dangerous signs can be immensely crucial in preventing such incidents from even happening in the first place. This will not only save countless human lives but also prevent the economic losses and slowdown that happens after the calamities occur. Taking small steps in the long term can help reduce the frequency and effect of such calamities to happen.

Satellite imagery can also be used to monitor private businesses and factories and the amount of smoke that they generate, and whether they are following the specified norms or not. This can be done by training models to detect smoke and estimate its intensity just by satellite images. This can be very useful in keeping a check on the pollution caused by these factories, especially in industrial areas where many such factories are located. This can be crucial in keeping the air quality in such industrial districts at normal levels, as such industrial districts suffer the worst from these polluters. It can also be used to identify the corporations that are breaking the law, making the job of the governing authorities far easier. Pollution from such factories affects their workers in the worst way, and they inhale many harmful gases hence this approach will also help keep many workers healthier and make their jobs easier on them.

CNN-based models can be trained to keep a check on deforestation and illegal cutting of trees. Deforestation is a major contributor to global warming and green and climate change. Deforestation not only contributes in a major way to climate change, but at the same time, it affects the animal life of the forest area and takes away the natural habitat of many animals. This forces the animals to leave the forests and travel into cities, causing much havoc and accidents. Several animals die and species go extinct only because of deforestation. It is estimated that on earth 137 species of animals go extinct every year because of deforestation alone. DL models can be trained to monitor forest caps and look for changes over a short and long time. This can give us crucial information on illegal deforestation activities and changes in forest caps in the long term. Action can be taken if the model shows changes in forest cap over a small time and the respective authorities in charge can be informed and action can be taken before the severity of the situation increases.

In the long term, DL models can give us a new perspective on the changes in forest cap over time. This can also be later on compared to the no of deforestation incidents that occurred during that time, again using DL models, and a relation can be established. DL models can also be used in monitoring ice caps, glaciers, and rivers. Models can be trained to observe long-term changes in ice caps and predictions can be done for the future. As glacier and ice cap melting is directly correlated to rising sea levels, which is yet another major sign of trouble that is about to come, the data regarding those is highly crucial. The melting of glaciers can be a very harmful thing. Some of the biggest cities around the globe are coastal cities that house millions of people and contribute significantly to global economies. Global trade relies on these coastal countries, and they are the ones which will face the worst impact due to the melting of glaciers and ice caps. Especially island cities that have a lot of coastal areas will be heavily impacted by this glacial melting. For example, the Maldives, an Island nation and a popular vacation destination for many travellers, located in the Indian Ocean, is predicted to lose 80% of its habitable land to rising sea levels by 2050. This is an alarming situation.

DL models can be trained to observe glacial melting patterns and make predictions. They can be trained to look for concerning signs and in general the overall changes in the ice caps. They can also be used to look at glacier movements in real-time and alert if something unusual is observed. They can also be trained to keep track of the ice cap and melting of the same. It can give us crucial data about the melting of ice caps and we can then predict its impact and prepare for the same. Another effect of climate change is the drying of rivers, caused due to global warming. Many rivers dry out in different parts of the world, causing water shortages. In some cases, rivers have over time changed their course due to climatic changes. DL models can be implemented to keep a check on the drying of rivers and the general long-term change in the pattern and the path rivers take. Satellite imagery can also be used to look after the cleanliness of the rivers. Models can be trained to identify unclean sections of the rivers, and the respective authorities can be alerted to act before the situation becomes worse. In many cases, many factories illegally dump waste. Another application of DL could be in monitoring soil patterns and changes in the soil. Changes in soil oftentimes reflect a climate change and can give out a lot of information about the soil beneath the visible surface level topsoil. Soil erosion is a major cause of concern in many parts of the world. It causes a lot of harm to the farming industry and in some cases, soil erosion can also lead to landslides. Soil erosion degrades the land and that makes the land less fertile. Hence, plants end up growing on the same land. This reduces the amount of carbon dioxide being absorbed by the plants in the same area. Soil erosion can also occur due to instant heavy rainfall, which is, in many cases, propelled by climate change. DL models can be trained to observe the changes in surface soil. They can be fed images of land plots just before the occurrence of landslides and hence they will be able to predict landslides before they happen.

These models can be fed with areas where severe soil erosion has already occurred, with historical photos of the same patch of land with healthy soil. This can help train the models for predicting soil erosion and land degradation and also label different patches of land based on their level of degradation and the possibility of occurrence of soil erosion. Models can be trained to alert the authorities if there is a high possibility of soil erosion or a landslide and the authorities and the landowners can then take the required action. This can also help agricultural researchers understand the soil and the impact of different farming practices on soil a lot better.

8.6 Other applications of this approach

There can be many other possible applications of this approach of using DL methods on satellite imagery, both in climate change and in other fields as well. A different application of DL in climate change monitoring could be long-term monitoring of monsoon data of different areas of cities and then linking it with satellite images and other satellite data for the same time dimension. This can be very helpful in many ways. It can be used in predicting floods based on satellite imagery. Models can be trained with pre-flood satellite images so that they could predict the occurrence of floods. This will end up saving countless lives. Floods have taken more than 100,00 lives from the year 1990 to 1999. Floods have killed at least 6.2 million people since the twentieth century. Floods cause huge economic and human capital losses. In some cases, floods have caused billions of dollars in losses. For example, the Chennai floods of 2015 caused 2.2 billion dollars in losses, while the Thailand floods of 2011 caused a humongous 30 billion dollars in losses.

Floods do not just cause damage to economies but cause massive displacement of people and in many cases, permanently damage the infrastructure of cities and towns. They disrupt the electricity supply to these places and the cell network also gets compromised. Years of work and money that go into building massive roads, flyovers, and public institutions, suddenly go in vain. While the suggested DL methods cannot prevent such events to happen, they can help us in predicting such events, so that early action can be taken. Many times, due to short notice, many people remain left to be evacuated from flood sites. If we have already predicted the occurrence of a flood well in advance, evacuation can start much early on and hence everyone can be safely evacuated. This also solves the problem of chaos that occurs at places where the evacuated people are taken. Such places can be prepared well in advance and hence much of the mismanagement that happens due to lack of time can be prevented.

With the proper implementation of DL models, such floods can be a lot less damaging and much more lives can be saved. Another application of DL models could be to monitor the desertification of land areas in the long-term using satellite imagery. Now, this approach is slightly different because of one major reason – desertification, unlike natural calamities, does not occur instantaneously, it occurs over a while. This means that a 24/7 feed on live satellite data is not needed, although it can certainly give a lot more data than it is needed to monitor desertification, this approach only requires a few snapshots every day. Previous research titled 'Status of Desertification in the Patagonian Region: Assessment and Mapping from Satellite Imagery', published by Argentinean researchers H.F. Del Valle, N.O. Elissalde, D.A. Gagliardini, and J. Milovich, explores the usage of satellite imagery in monitoring desertification of an area called 'Patagonian Region'. They found that of the total region studied (78.5 million ha), 93.6% (73.5 million ha) showed different degrees of desertification. Categories of desertification for the whole region were slight (9.3%), moderate (17.1%), moderate to severe (35.4%), severe (23.3%), and very severe (8.5%). This paper was published in 1997 and the measurements were done manually. And while no DL/ML algorithms were used here, it just demonstrates the ability of satellite imagery and its potential just by itself.

If DL models are trained to identify areas that are going through desertification and also train the models to determine the severity of desertification and give them a rating, then this opens up a lot of possibilities. Areas with controllable severity levels can be saved. Local authorities can be informed, and action can be taken. Moreover, models can be trained to predict the speed of desertification and its spread, and potential areas where desertification can start to happen. This can be used to alert the stakeholders to take action. This data can also be very useful to corporations and government projects, where teams look for land for different projects. This can help projects where a very specific type of land is required. Droughts have also become a significant challenge in many countries around the globe, and droughts can lead to the desertification of lands. Observing the growth and spread patterns of droughts using satellite images can also help us immensely. Models can be trained with images of pre-drought land so that the possibility of the occurrence of droughts can be predicted, and hence the required action can be taken.

Droughts take countless lives every year around the globe in many different countries. They cause a lack of water, any kind of water, be it drinking or utilitarian water. This leads to countless deaths. Droughts primarily affect relatively poorer people, and it is the poor section of society that faces the worst. Predicting droughts can enable authorities to act accordingly to prevent such dire situations from happening.

8.7   Possible problems

While the idea of applying DL techniques to satellite imagery to monitor climate change and its impacts is very efficient and can be extremely helpful during evacuation and rescue missions, it has its share of issues and problems that may make the process, at times, much less efficient. One such issue is the requirement of specialized and powerful systems required to run the programs. When the entire globe's satellite feed is being monitored using advanced deep-learning models, the process is bound to be resource-heavy. Not every place may have the required systems that are powerful enough to run such programs.

Another issue is false positives. While the error rates of the discussed models are fairly low, when it comes to natural disasters, even one false alert about a calamity can cause unnecessary panic. Hence, the alerts must be properly scrutinized before the authorities are informed. This requires the availability of personnel who can be present to scrutinize the alerts and inform the respective authorities. And this again requires people with a technical understanding of the program, and such people may not be present in certain parts of the world. Another possible issue could be slow communication with the authorities. While the program may have alerted the people handling the program if the communication systems have been damaged, or if there is any delay in alerting the respective authorities, then even though the program worked as expected, the delay will lead to reduced efficiency and the intended results may not be achieved. And while not a direct issue with the models, it certainly can reduce the impact and the intended results may not be achieved, and even end up risking some lives.

Another downside of this approach is that it heavily relies on live satellite imagery. If due to some unforeseen reasons, a certain satellite breaks down or stops sending signals back, then this approach cannot work. It needs live satellite imagery to detect calamities.

8.8   Conclusion

Climate change has become a majorly discussed issue around the globe. It has become one of the main causes of deaths and economic losses worldwide. CIDs have increased in frequency in the past few years and action needs to be taken to reverse the issue of climate change. The first step in curbing climate change is to effectively monitor it and take instant action at the time of a calamity. Monitoring the impacts of climate change will be crucial in tackling it, and reducing the consequences and losses caused due to disasters. Active and efficient monitoring of disasters can help in early warning, which can help in a faster and more effective response, saving countless lives and reducing losses.

Satellite imagery plays a crucial role in monitoring climate change and observing its impacts around the world. Wildfires, landslides, volcanoes, etc. can be identified and monitored with satellite imagery. Satellite imagery has historically been used to monitor the long-term effects of climate change, like changes in the ice cap, forest cover, etc. Satellite images of volcanic eruptions, wildfires, landslides, etc. help in mapping the data resulting in more efficient rescue operations. Modern DL methods can be extremely useful in assisting the monitoring of climate change using satellite imagery.

The idea of applying DL methods to satellite images for monitoring climate change and its impacts has been proven to be highly effective and efficient in both identifying and predicting the spread of natural calamities and can be of great use to the authorities responsible for rescue and evacuation missions. The DL methods discussed here have the ability to not only identify the happening of the calamity but also to localize and find out the exact region that is being affected. Over time, some models can learn and begin to predict the probability of the occurrence of a climate-induced disaster. The usage of U-Net, a DNN model initially introduced as a bio-medical image segmentation model, in land segmentation for detecting patches of land burnt after a wildfire, shows the possibilities that DL provides.

The use of modern DNNs is much more efficient than manual monitoring done by humans. It cuts down on labour costs, reduces the time taken, increases the efficiency of disaster detection, and is even capable of predicting the occurrence of such events. This ends up saving much more lives and reduces economic losses. It also helps the rescue authorities in preparing for such an event in advance. Hence, this is an extremely useful approach and if implemented this will save many lives and reduce economic losses when calamities happen. The implementation of these DL methods and CNN models requires far less capital investment and requires a lot less personnel. It also keeps improving itself as time goes by and requires very little change.

This approach has also been proven to be extremely accurate and reliable and is much faster than conventional methods. Hence, this approach makes a very strong case for itself and can be extremely useful not just in climate change monitoring, but at the same time, it can also in helping save countless lives at the time of CIDs and natural calamities.

References

[1] Callendar, G.S. 'The artificial production of carbon dioxide and its influence on temperature', *Q. J. R. Meteorol. Soc.*, 1938, **64**(275), pp. 223–240.

[2] 'A brief history of climate change', BBC News, 2013.

[3] 'COP26: Together for our planet | United Nations', https://www.un.org/en/climatechange/cop26, accessed March 2022.

[4] 'UN Climate Change Conference (COP26) at the SEC – Glasgow 2021', https://ukcop26.org/, accessed March 2022.

[5] 'Evidence | Facts – Climate Change: Vital Signs of the Planet', https://climate.nasa.gov/evidence/, accessed March 2022.

[6] Forcada, J. and Trathan, P.N. 'Penguin responses to climate change in the Southern Ocean', *Glob. Change Biol.*, 2009, **15**(7), pp. 1618–1630.

[7] Veytia, D. and Corney, S. 'Climate change threatens Antarctic krill and the sea life that depends on it', http://theconversation.com/climate-change-threatens-antarctic-krill-and-the-sea-life-that-depends-on-it-138436, accessed March 2022.

[8] 'The Global Risks Report 2020', https://www.weforum.org/reports/the-global-risks-report-2020/, accessed March 2022.

[9] Jones, M.W., Smith, A.J.P., Betts, R., Canadell, J.G., Prentice, I.C., and Le Quéré, C. 'Climate change increases the risk of wildfires: January 2020', in: *Science Brief*, 2020.

[10] Ban, Y., Zhang, P., Nascetti, A., Bevington, A., and Wulder, M. 'Near real-time wildfire progression monitoring with sentinel-1 SAR time series and deep learning' *Sci. Rep.*, 2020, **10**, p. 1322.

[11] Rolnick, D., Donti, P.L., Kaack, L.H., *et al.* 'Tackling climate change with machine learning', 2019, ArXiv190605433 Cs Stat.

[12] Khryashchev, V. and Larionov, R. 'Wildfire segmentation on satellite images using deep learning', in: *2020 Moscow Workshop on Electronic and Networking Technologies (MWENT)*, 2020, pp. 1–5.

[13] Di Napoli, M., Marsiglia, P., Di Martire, D., Ramondini, M., Ullo, S.L., and Calcaterra, D. 'Landslide susceptibility assessment of wildfire burnt areas through earth-observation techniques and a machine learning-based approach', *Remote Sens.*, 2020, **12**(15), p. 2505.

[14] Francis, P.W., Wadge, G., and Mouginis-Mark, P.J. 'Satellite monitoring of volcanoes', in: Scarpa, R. and Tilling, R.I. (Eds.), *'Monitoring and Mitigation of Volcano Hazards'*, Springer, 1996, pp. 257–298

[15] Segall, P. 'Volcano deformation and eruption forecasting', *Geol. Soc. Lond. Spec. Publ.*, 2013, **380**(1), pp. 85h–106.

[16] Pepe, A. and Calò, F. 'A review of interferometric synthetic aperture RADAR (InSAR) multi-track approaches for the retrieval of earth's surface displacements', *Appl. Sci.*, 2017, **7**(12), p. 1264.

[17] Anantrasirichai, N., Biggs, J., Albino, F., and Bull, D. 'A deep learning approach to detecting volcano deformation from satellite imagery using synthetic datasets', 2019, ArXiv190507286 Cs Eess.

[18] Shakya, S., Kumar, S., and Goswami, M. 'Deep learning algorithm for satellite imaging based cyclone detection', *IEEE J. Sel. Top. Appl. Earth Obs. Remote Sens.*, 2020, **13**, pp. 827–839.

[19] Priya, R.S. and Vani, K. 'Deep learning based forest fire classification and detection in satellite images', in: *2019 11th International Conference on Advanced Computing (ICoAC)*, 2019, pp. 61–65.

[20] Lv, H. and Chen, X. 'Research and implementation of forest fire smoke detection based on resnet transfer learning', in: *Proceedings of the 2021 5th International Conference on Electronic Information Technology and Computer Engineering*, Association for Computing Machinery, 2021, pp. 630–635.

[21] Krizhevsky, A., Sutskever, I., and Hinton, G.E. 'Imagenet classification with deep convolutional neural networks', in: *Advances in Neural Information Processing Systems*, Curran Associates, Inc., 2012.

[22] Szegedy, C., Vanhoucke, V., Ioffe, S., Shlens, J., and Wojna, Z. 'Rethinking the inception architecture for computer vision', 2015, *ArXiv151200567 Cs*.

[23]   Iandola, F.N., Han, S., Moskewicz, M.W., Ashraf, K., Dally, W.J., and Keutzer, K. 'SqueezeNet: AlexNet-level accuracy with 50x fewer parameters and <0.5 MB model size', 2016, ArXiv160207360 Cs.

[24]   He, K., Zhang, X., Ren, S., and Sun, J. 'Deep residual learning for image recognition', 2015, ArXiv151203385 Cs.

Chapter 9

A comparative study on torrential slide shortcoming zones and causative factors using machine learning techniques: a case study of an Indian state

G. Bhargavi[1] and J. Arunnehru[1]

Landslides are a dangerous geomorphological phenomenon frequently occurring in mountain terrain. Landslides cause many geomorphological inclinations, including rockslides, slope recession, and minor debris continuing to flow. Typically, slope failures are without prior notice. It provokes significant damage throughout roads, railways and connecting infrastructures, affecting all modes of public transport, civilised residential apartments, cultivating farms, valleys, dense forests, and so on. This geomorphological transformation may result in massive loss of people's livelihoods and belongings, mineral wealth critical to the nation's economy, and economic development depletion in urban areas. However, one of the primary causes of the mudslide occurrence is the financial diminishment of the peak district. The probability of catastrophic landslides over the Western Ghats' downwind slope during heavy downpours relates to the region's particular geomorphology and environment. According to Kerala's peculiar geomorphological environment, the southwestern monsoons' annual precipitation causes catastrophic landslides. As a result, the latest research was carried out to investigate the landforms of the terrain in-depth, which seems dominated by tectonic plates, to determine its influence on landslides. This chapter also attempts to learn more about landslides and the various factors that influence landslides that suit Indian terrain and environmental conditions.

9.1 Introduction

Landslide susceptibility is heavily influenced by topography. Failures can happen on the coastline, in the highlands, in the Midlands, and near shore. The plain landscape has low gravitational forces, which makes it the ideal factor for disasters [1]. Kerala is a tropical climatic region in India's southwest. Landslides are common in this state due to the torrential rains and the persistent clay soil. It covers an area of 38,863 square

[1]Department of Computer Science and Engineering, SRM Institute of Science and Technology, India

kilometres. The monsoon serves up heavy cloudbursts to the massively influenced region on an annual basis. It receives 3,107 mm on an average [2]. When compared to neighbouring states, the variation in annual rainfall is minimal. However, there have been chronicler moisture years observed, such as 1924, which received 3,368 mm of rain. The year 2018 saw 310 mm of rainfall in 48 h. The primary purpose is to quantify the risk of landslides in Kerala, precisely characterising landslides caused by rain. Rainfall is among the most critical factors that cause landslides, and it is one of the most devastating naturally occurring disaster events in the atmosphere [3]. The Idukki district started to receive 36% more thundershowers than average, causing widespread flood events and pervasive flash flooding and mercilessly slaughtering 445 residents. During the August 2018 major storm catastrophic event, water-covered areas increased by almost 90%. A variety of factors, including inclination, characteristics, height, slope, contours, land nature, proximity from sewage treatment, land use, and land cover, as well as a satellite image, are included. It is essential to recognize susceptible landslide locations to avoid additional damage [4].

Many researchers should analyse the geomorphological and climatic changes to extract a patterned and reliable prediction event of landslides. Omid Ghorbanzadeh [5] gives a brief explanation of landslide detection in Himalayan ranges. They are using optical data from satellite images. Using machine learning algorithms author generates various triggering factor maps to detect and identify the landslide in the study area. But they failed to segregate the human settlement in the study area which is considered the most important objective of the research. Devara Meghanadh [6] employs the analytical hierarchic process (AHP) to evaluate landslide susceptibility, which determinates the rating of each triggering factor of disaster areas identified in Landsat images and evaluates landslide vulnerability from the overall factors.

Faming Huang [7] and his team identify landslides in forest areas in China. They tried various machine learning algorithms in which they built a model using a random forest algorithm and support vector machine with DTM results based on object-oriented strategies.

The existing landslide research does not give clarity about the landslide-triggering factors and the severity of the disaster. Although they have used various techniques to identify the landslide causes, there is a humongous gap that is found between landslide vulnerability mapping and landslide risk assessment. To begin the research, the weak spot should be evaluated by evaluating the spatial and temporal value of the landslide. The objective is to split landslide data from publicly available accurate documents. The evaluation was planned to explore original information as a cost-effective way for landslide inventory planning to furnish workable solutions to the determined challenges using various machine-learning approaches. The objective is to split landslide data from publicly available accurate documents (Figure 9.1).

As a result, administrators are collaborating with teams to investigate landslides and mitigate their effects. Researchers paid close attention to research fields such as landslide risk, known vulnerability zones, disaster preparedness, modelling techniques, sloped area monitoring, geohazard, and vulnerability assessments.

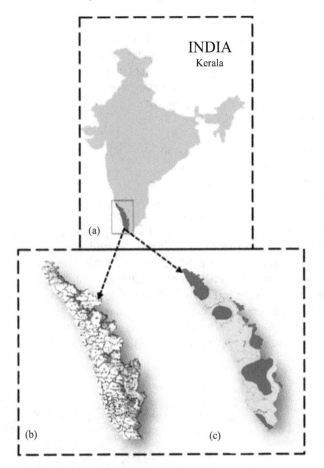

Figure 9.1   (a) Map of India highlighting the state Kerala. (b) Highlights the landslide zones and (c) highlights the 2021 annual rainfall data.

Landslide mapping employs a variety of methodologies, artificial intelligence techniques, and open-source GIS software [8]. This disaster leaves a distinct imprint on the Earth's atmosphere, meticulously recorded at regular intervals by earth observation satellites or remotely sensed data. Modern disaster relief activities rely on satellite data to identify risk areas quickly, provide emergency relief to affected areas, assess post-disaster management, and accurately identify victims. Data are freely available from various remote sensing data hubs.

9.2   Discussions on landslide influencing factors

The creation of the landslide-impacted region as landslide mapping and classification of the detailed landslide risk area is the most significant contribution of this research. This sensing and monitoring can assess the danger of a disaster, and an alarm system can be created as a result. Even though we can predict landslide frequency, the Google Earth engine datasets will define the landslide vulnerability zone [9]. As a result, monitoring and forecast zones will almost certainly be established. The goal is to improve the centralised public database by employing satellite photographs and accompanying data to aid disaster response by pinpointing specific regions – the ability to develop accurate landslide risk zonings for large areas using verifiable approaches. Furthermore, as confirmed by a survey done with the relevant research publications, machine learning approaches such as regression models and support vector machines (SVM) were employed to develop accurate landslide predicting models noting that the Western Ghats of Kerala are prone to repeated landslides as a result of complex tectonic geomorphology and significant precipitation penetration, resulting in increased pore pressure and landslides [10]. Deforestation has also made it easier for rainwater to enter directly. The preceding investigations corroborate our hypothesis that landslides are caused by pore pressure. Thus, avalanche peril planning is an initial step that will incorporate the information we wanted to settle on choices about the affected region and distinguish inclinations for proper relief arrangements [11]. However, these are practically equivalent, they are unique concerning the mark of cycles and the ideas. Since the current review has zeroed in additional on geomorphology in the avalanche-inclined regions, the wording 'inclination' has been utilized. Figure 9.2 shows the

Figure 9.2   Viewpoint on Idukki torrential slide and adjoining areas. The dim square shapes put aside in the figure show minor torrential slides occurred in a comparable whirlwind event close by, and the yellow square shape on the upper right of the figure exhibits the space of the breaks.

aerial view of a landslide that occurred in the year 2018 in the Idukki district of Kerala. The landslide expanded up to 1,200 m from the peak.

9.3   Materials and methods

Computerized height models (DEMs), spaceborne engineered gap radar (SAR), multi-worldly LiDAR symbolism, optical remote detecting pictures, and site studying estimations have all been effectively utilized. The data was assembled by the as of late dispatched Sentinel-2 satellite. The most prominent trait of the Sentinel-2 satellite is its 5-day return to the Equator [12]. In sans cloud circumstances, the spatial goal of a satellite fluctuates up to 10 m for various recurrence groups or stations, for example, Band 2 – blue, Band 3 – green, Band 4 – red, and Band 8 – infrared. We can join at least two groups for our execution because each band has exceptional properties. The system that utilizes many groups produces comparative discoveries that demonstrate relative patterns. These outcomes were made by broadening the picture into mathematical geological representation utilizing the digital elevation model (DEM). DEM was made by the Indian Remote Sensing Satellite (IRS) [13]. DEMs are utilized to make geographic qualities like slants at some random position, tendency, and look. A DEM is a three-dimensional image of a landscape surface. Printing shape lines are changed into DEMs (Active form approaches in DIP). Polygons are made utilising specific computerized limits, and every polygon follows the height information from the limit layout before being utilised as a raster or vector information design. DEMs are fundamentally liable for GIS appropriateness. Changes to the DEM can be made to gauge disintegration, precipice disappointment, and decide avalanche volume.

9.4   Dataset collections

Satellite images are the primary source of data for landslide forecasting. The number and quality of images captured by satellite remote sensing are rapidly increasing. It is essential for monitoring the Earth's surface. For a better understanding of the scenario, satellite images, field survey data, environmental data, and historical data were used in conjunction with satellite images. To extract meaningful data from raw satellite images, the ArcGIS tool is used. These landslides are difficult to identify from optical remote sensing data due to dense vegetation cover and snow cover. DEM data are used to reveal precise morphological changes such as altitude, slope, and slope aspect. Many advanced DEM data are publicly available on the USGS website.

9.5   Rainfall characteristics in Kerala

DEMs, spaceborne engineered gap radar (SAR), multi-worldly LiDAR symbolism, optical remote detecting pictures, and site studying estimations have all been effectively utilized. The data was assembled by the as of late dispatched Sentinel-2 satellite [14]. The most prominent trait of the Sentinel-2 satellite is its 5-day return to the Equator. In sans cloud circumstances, the spatial goal of a satellite fluctuates up to 10 m for various recurrence groups or stations, for example, Band 2 – blue, Band 3 – green, Band 4 – red, and Band 8 – infrared. We can join at least two groups for our execution because each band has exceptional properties [15]. The system that utilises many groups produces comparative discoveries that demonstrate relative patterns. These outcomes were made by broadening the picture into mathematical geological representation utilizing the DEM. DEM was made by the IRS satellite. DEMs are utilised to make geographic qualities like slants at some random position, tendency, and look. A DEM is a three-dimensional image of a landscape surface. Printing shape lines are changed into DEMs (active form approaches in DIP). Polygons are made utilising specific computerized limits, and every polygon follows the height information from the limit layout before being utilised as a raster or vector information design [16]. DEMs are fundamentally liable for GIS appropriateness. Changes to the DEM can be made to gauge disintegration, precipice disappointment, and decide avalanche volume.

Precipitation and the timetable time frame for the locale repository are other key setting-off factors for the avalanche. Precipitation, as per [17], is an occasional impacting factor with a stage-like edge bend. Other affecting components, for example, tremors and anthropogenic exercises produce an undulation shape in the edge bend.

Avalanche identification is portrayed [18]. This distinguishing proof is fundamental and addresses a stage forward during the time spent fast danger evaluation and relief. The avalanche weakness region is determined utilising numerous open-source methods utilising avalanche planning. The ghastly properties of the avalanche regions are utilised to group them. Picture division is utilised to distinguish the change discovery, trailed by the expulsion of non-helpless areas utilising an item-based methodology. At last, utilising the solo strategy to bunch the picture protests, the yield is recovered and characterized [19]. The splendour of the creation of the post-avalanche picture recognises the pre and post-avalanche satellite photos. From the avalanche planning, the post-avalanche edge esteem shows the avalanche-harmed region. Likewise, avalanche separating limitations, for example, varieties in the DEM, slant, principal component analysis (PCA), and Green Normalized Difference Vegetation Index (GNDVI) evaluations were utilized to diminish invigorating sham expectations. Utilising the k-implies bunching calculation, the things that stay after counterfeit asset decrease are isolated into two gatherings. By setting up the principal characteristics, the topographical attributes included with a model can be approximated. Figure 9.3 portrays the rainfall range that occurred in various districts in Kerala for researching avalanche subtleties. The parkway street between Madurai and Munnar was enlarged because of an avalanche that occurred in 2021. The avalanche stock is made utilising high-goal satellite information from before and then afterwards the avalanche [20]. For avalanche planning, an assortment of uninhibitedly accessible satellite pictures is utilised.

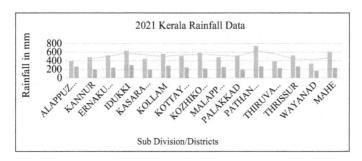

Figure 9.3   2021 Kerala rainfall data from Indian Metrological Department, Thiruvananthapuram

As per Hasali Hemasinghe *et al.*, landslide helplessness is characterised as a future pattern toward an avalanche in a particular region. This is a computation of the connection between the deciding specialists dependent on the geological example of their tendencies. Avalanche helplessness planning distinguishes avalanche areas of interest around the country dependent on a bunch of actual precipitation conditions. It is known as landslide susceptibility zonation, which isolates the scope of the land cover into expected comparable zones and positions them as plausible avalanche peril zones. The GIS-based slope unit technique is utilised by Baeza. When deciphering an avalanche stock, the letter k indicates a bunch that isolates the region into avalanches and non-avalanches. Autonomous examples with a similar example measurement are picked indiscriminately as preparing information for the avalanche-determining model. The excess examples were picked as gauging or testing to guarantee that the avalanche figure idea was right [21]. Avalanche spatial estimating depends on grouping task translation. This is a standard technique utilised in avalanche defenselessness planning depending on precipitation. The discriminant examination was utilised to distinguish the evaluation of the discriminant work condition and to isolate the avalanche-inclined and non-inclined zones. The information should be parted into touchy and non-helpless regions to prepare the avalanche expectation models. The powerlessness to risks differs from low to high to unimaginably high. Seismic movement and substantial deluges are the most widely recognised events in the area. Along these lines, the circumstance gets progressively risky as anthropogenic demonstrations and climate varieties increment. The greatest avalanches in this space are in a general sense driven by precipitation, according to verified evidence.

Abhirup Dikshit *et al.* give a short outline of the phases of avalanches and the components that impact their event. Discovery, forecast, and example arrangement are fundamental stages in avalanche requests previously, then after the fact of the event. Researchers have endeavoured to depict avalanches and their connected repercussions utilising sound system picture handling because of progressions in geology and simple admittance to high spatial and multi-transient satellite pictures. Picture from previously, then after the fact the avalanche landslide distinguishing proof procedures incorporate change the identification, picture coordinating, and mechanised DEM combination [22]. As a rule, there is an assortment of picture arrangement procedures. When used for high-goal photographs, the pixel-based ID strategy has a few restrictions. The utilisation of an article as opposed to a pixel is utilised to pass on depictions. It adds to our cognizance of ordinary misfortunes, for example, torrential slides because the misfortunes are of remarkable size and look. Shape, logical, spatial, and ghostly provisions are completely utilised in object-based methodologies. A quantitative examination can be adequately inferred utilizing the DEM.

9.6   Landslide impacted earthquake

The ground moves during a tremor, compelling frail grades to disintegrate. At the point when a slanted substance becomes wet because of a downpour, it may bring about a landslide or wreck stream. Post-quake tremors of the greatness of 4.0 or higher ought to be identified with torrential slides. Trees, structures, and vehicles might be evacuated because of the stone and mud flotsam and jetsam; accordingly, coasting materials might obstruct extensions and streams, making floods along the way [23]. Managers and organizers can utilise avalanche hazard guides to see precisely where this danger should be evaluated before improvement in high-hazard regions should be possible securely.

In India, avalanches brought about by quakes are recorded at both the large-scale and miniature levels in the Himalayan area and the Western Ghats locale. Because of its calm environment, India, similar to some other nations, has little effect. As indicated by the National Disaster Management Authority, around 15% of all our land use is affected by the chance of avalanches National Disaster Mitigation Agency (NDMA). Avalanches are an issue in Karnataka's sloping regions; the consolidated effect of seismic earth-shaking and precipitation will bring about wide-reach defoliation [24]. Tremor-incited avalanches are generally described as the earthward and evident vibration of slant materials, for example, rock or soil in the slope area under the impact of gravity, as announced by the NDMA.

Seismic tremor-instigated avalanches were surveyed utilising GIS strategies by Naveen James *et al*. He pronounces perceiving the direct source design; a probabilistic seismic danger appraisal was led to anticipate top-level speed increase at the stone bed. The territory point at every network point ought to be more huge than 10°. The inert portrayal of security demonstrated to keep up with the avalanche was utilised to register the Tectonic avalanche hazard to every surface network. Utilising GIS, an incline map, is made to decide the slanted point of all lattice focuses. For every matrix point, the inert portrayal of the slip security required can be determined [25]. The dormant component of safety related to a tendency decides protection from quakes brought about by that tendency. The tendency worth ought to be a basic static component for insurance to exhibit that the region is more steady than seismic tremor vibration. Trim tension is applied to the sliding degree because of the seismic tremor's steady filling, and trim powers can defeat trim unsusceptibility.

As a methodology for examining seismic tremor-actuated avalanche situations, Martino *et al.* propose the Probabilistic Approach pRovide Scenarios of seismic tremor Induced incline FAiLures (PARSIFAL). The PARSIFAL model glances at the recurrence of starter avalanches and miniature earthquakes [26]. Miniature level breakdown happens first on slants, unaffected by motions. Then again, a repeat of this miniature level might advance to a full-scale level, which might be affected by structural plate development. Over-the-top precipitation, which triggers both surface developments, has been seen because of outside or inward factors like tremor action and opening tensions. The susceptivity to repeat is assessed by distinguishing and auditing the beforehand existing avalanches utilising GIS.

9.7   Anthropogenic activities

Stream remaking, legitimate and illicit mining, foundation maturing, territory change, land-use transformations, street and rail route advancement, deforestation, and other anthropogenic factors are on the whole expanding avalanche risks [27]. Expressway and rail line improvement regularly includes converging slants and eliminating material from slopes. Trees are being felled to extend the street, yet there is no soil to exhume. Expressways at the middle and lower part of the slope address the main avalanche hazard because of water obstruction. As indicated by a new report led by specialists at the University of Sheffield in the United Kingdom, avalanches killed an expected 50,000 individuals each year worldwide somewhere in the range between 2004 and 2016. Around 20% of those occasions happen in India, as indicated by gauges [28]. As indicated by Avinash *et al.*, anthropogenic exercises produce high defenselessness avalanches in the upper districts of the Western Ghats region. It is likewise where human-caused avalanches are quickly spreading, influencing Sri Lanka, Pakistan, Nepal, Myanmar, and other adjoining nations.

'Torrential slides set off by cutting trees on inclines are typically an obstruction in rustic zones, where numerous clans and individuals unlawfully take wooden sticks and other unrefined substances off the slope to develop their homes' [29] noted. Martin Haigh and Jiwan Singh Rawat led a contextual analysis in the Almora region of Uttarakhand. As per him, the current development of metropolitan districts has a huge effect. In 2010, there were just nine avalanches on Kilbury Road yet 108 avalanches in the Almora Lower Mall locale. Avalanches should be inspected at various geographic and natural frequencies, as per Canuti *et al.* [30]. The utilisation of remote detecting methods for anticipating avalanche constraint and reconnaissance is immature, even though specific methodologies take into account quicker information securing across enormous regions and fill in as a basic apparatus for cautious avalanche control.

Avalanche planning is a fundamental basic perspective in assessing the recurrence of avalanches. The social and financial meaning of avalanche disintegration is portrayed through stock information. Satellites would now be able to do a quantitative appraisal of landscape improvements over a wide geographic reach in scantily vegetated regions. Utilising a SAR satellite might work on the current technique for noticing incline movement at a geological split. For avalanche observing or unexpected tendencies, earthbound or ground-based SAR interferometry framework strategies are an urgent instrument. The joined presentation of the sensors and territory activities, while thinking about projected headways, can fundamentally assist with a few pieces of information and observing limitations related to the avalanche pointer.

9.8   Machine learning techniques for landslide study using satellite images

9.8.1   *Highlights of machine learning techniques in satellite images*

Satellite imagery is one area where machine learning is particularly important. It aids in addressing the global issue that ultimately has an impact on people's lives and a nation's socioeconomic issues. Many different types of raw data are gathered by satellite images all over the world, but these data can only be partially accessed and are subject to manipulation. If these data are combined with machine learning techniques, so many complex global challenges can be addressed. The data made available to researchers and data scientists will bring about changes in global complex challenges such as changes in the climate, forest fires, changes in land use and land cover over time, ecology imbalance, and drastic changes in population, as well as improve a country's economy. Large data sets that are accessible, real-time monitoring, simple and accurate prediction methods, and profound cost-effectiveness are the main benefits of machine learning. An enormous amount of raw, unstructured satellite data is effectively transformed into structured, meaningful information when satellite images and machine learning algorithms are combined. This combination is now in high demand, and many civil engineers and government officials specialise in this research area to simplify and improve the accuracy of a complex task.

In satellite images, profound learning can be used in three potential ways: grouping of surfaces, changing location, and item discovery. Recent survey results show that all specific issues are deciphered as an item and change identification tasks that are dealt with by utilising AI strategies to achieve best-in-class results based on computer vision, to improve and acclimate the objectives of remote detecting-based applications, overcoming unambiguous hindrances [31]. Landslide researchers have evidenced that machine learning and its derived products, such as deep learning (DL), can successfully be employed in landslide-related analyses. This has been shown in numerous previous research, and it solves image classification problems accurately.

All AI approaches have been used moderately for dissecting the potential danger of avalanches, and it heavily relies on stock datasets of the review region with the known spatial extent of avalanches. When applying machine learning techniques to a large number of datasets for preparation and approval, not many advances are required. It is critical to identify and design avalanche-influenced regions to support planning and emergency response on time [32]. The most noticeable techniques are Decision Tree, ANN, SVM, and gathering strategies like Random Timberland and Bagging. Early AI approaches are used with decision trees, SVM, and neural network centre calculations. Current research is overwhelmingly focused on the improved variant of ML calculations and their subsidiary or mixture structures [33].

The above-mentioned improved versions are used for avalanche powerlessness investigation, avalanche removal forecast, determination of relevant moulding factors, and avalanche region. The capability of SVM, DT, NN, and RF with GIS datasets and remote detecting images for avalanche powerlessness planning has been investigated. RF has gotten a lot of attention recently because of the following advantages: incredible precision, the fastest handling speed, and the ability to research high-dimensional data. They promoted a hybrid approach, known as the 'Multi support Based Nave Bayes Tree', to predict the spatial extent of avalanches [16]. Table 9.1 explains the various machine learning algorithms and their purpose.

Table 9.1   List of algorithms for landslide susceptibility vulnerability analysis

Algorithm	Purpose	Implementation	Others
Random forest – tree based	Regression, classification, feature selection	Two random processes at work: (i) Bootstrap algorithm is used to select training samples, and (ii) random sampling of feature subsets using tree induction	–
Decision tree	Representation of a structural pattern in data that does not have a relationship with the input variable to the objective parameter	The hierarchical model, which includes nodes, subtrees, and branches	Disadvantage: various outputs are not permitted, and the system is vulnerable to imbalanced datasets. Advantage: it is understandable and organised
Boosting	Switch a weak learner into a strong learner	In the series, construct by utilising vulnerable learners	AdaBoost is an effective algorithm for boosting
SVM – Kernel based	Optimal hyperplane finding	Kernel functions like linear, sigmoid, polynomial, and radial basis functions are employed	The radial basis function is mainly used for landslide modelling
Neural network based on backpropagation	Chain rule training is used to train the neural network	Neural network with multiple layers	With fewer computer operations, you can achieve high prediction accuracy

9.9   Emergency rescue and mitigation

The incline should be explored and checked on to anticipate their presentation and look further into their opposition, well-being, soundness, and disfigurements [34]. The opposition to the tendency is assessed by utilising different GIS programming and factual reflection. The investigation of slant opposition is turning out to be progressively significant in the field of geospatial designing. The grouping of parts in danger, like individuals, designs, structures, and framework in the district, vegetation property, or natural conditions – existing items and the region under peril defenselessness zone later on – is the initial phase in the avalanche hazard evaluation process [35]. The convergence of the peril power map and the guide of a component in danger, where geological and transient weakness esteems are considered, is utilised to survey the danger. Tapas R. Martha [36] examines how semi-computerised strategies were utilised to make avalanche inventories from satellite pictures taken after avalanche scenes. It is utilised to appraise avalanche defenselessness, hazard, and peril in India's hillslope locale.

9.10   Conclusion

This audit paper plans to understand the viewpoints that impact avalanches in India's meteorological conditions. Avalanches are a typical normal disaster, especially in the Himalayan and Western Ghats areas. As per this review, water and stressed water after precipitation from an adjoining state assume a basic part of Indian avalanche mishaps. We discussed the course of avalanche occasions since planning avalanches is the initial step. Then, at that point, it is separated into steps like estimation, checking, demonstrating, and at last danger appraisal and alleviation, which is alluded to as the board. AI calculations and remote detecting information can propel this work and give an authentic contribution as a powerful occasion or catastrophe caution. We can likewise make an early admonition model to forestall the annihilation. We can download an assortment of time-series picture information to comprehend avalanche highlights in a specific region better. Outrageous environmental changes will be knowledgeable about the future, inciting the advancement of information put together innovation based on past information utilising profound learning. The public authority can give assets for compelling alleviation activities because of the phenomenal advancement in satellite photographs, which permits them to detect the lowered zones and sort out salvage tasks all the more rapidly. Besides, the exploration exertion will be useful to an overseer, engineers in arranging, and specialised specialists in making financial exercises in such an area.

References

[1]   P.P. Sun, M.S. Zhang, and L.F. Zhu, "Review of 'workshop on landslide in Southeast Asia: management of a prominent geohazard' and its enlightenment," *Northwest. Geol.*, 2013.

[2] P. Preji and B. Longhinos, "Landslide susceptibility analysis and mapping in Sastha valley of Periyar River Basin," in *Lecture Notes in Civil Engineering*, 2021, vol. 86, pp. 23–41, doi: 10.1007/978-981-15-6233-4_3.

[3] S. Joseph, "Preliminary analysis of a catastrophic landslide event on 6 August 2020 at Pettimudi, Kerala State, India," *India Landslides*, vol. 18, pp. 1459–1463, 2021, doi: 10.1007/s10346-020-01598-x.

[4] A. Rajaneesh, B. Brototi, K. S. Vignesh, *et al.*, "Landslide susceptibility mapping using integrated approach of multi-criteria and geospatial techniques at Nilgiris district of India," *Arab. J. Geosci.*, vol. 14, pp. 1553–1568, 2021, doi: 10.1007/s00254-008-1431-9.

[5] O. Ghorbanzadeh, T. Blaschke, K. Gholamnia, S. R. Meena, D. Tiede, and J. Aryal, "Evaluation of different machine learning methods and deep-learning convolutional neural networks for landslide detection," *Remote Sens.*, vol. 11, no. 2, pp. 1–21, 2019. doi: 10.3390/rs11020196.

[6] D. Meghanadh, V. Kumar Maurya, A. Tiwari, and R. Dwivedi, "A multi-criteria landslide susceptibility mapping using deep multi-layer perceptron network: a case study of Srinagar-Rudraprayag region (India)," *Adv. Sp. Res.*, vol. 69, no. 4, pp. 1883–1893, 2022, doi: 10.1016/J.ASR.2021.10.021.

[7] F. Huang, J. Yan, X. Fan, *et al.*, "Uncertainty pattern in landslide susceptibility prediction modelling: effects of different landslide boundaries and spatial shape expressions," *Geosci. Front.*, vol. 13, no. 2, p. 101317, 2022, doi: 10.1016/J.GSF.2021.101317.

[8] L. Hao, A. Rajaneesh, C. van Westen, *et al.*, "Constructing a complete landslide inventory dataset for the 2018 monsoon disaster in Kerala, India, for land use change analysis," *Earth Syst. Sci. Data Discuss.*, vol. 2, pp. 1–32, 2020, doi: 10.5194/essd-2020-83.

[9] J. Jacinth Jennifer and S. Saravanan, "Artificial neural network and sensitivity analysis in the landslide susceptibility mapping of Idukki district, India," *Geocarto Int.*, vol. 37 no. 19, pp. 1–24, 2021, doi: 10.1080/10106049.2021.1923831.

[10] H.R. Pourghasemi, A.G. Jirandeh, B. Pradhan, C. Xu, and C. Gokceoglu, "Landslide susceptibility mapping using support vector machine and GIS at the Golestan province, Iran," *J. Earth Syst. Sci.*, vol. 122, no. 2, pp. 349–369, 2013, doi: 10.1007/s12040-013-0282-2.

[11] A.L. Achu, C.D. Aju, and R. Reghunath, "Spatial modelling of shallow landslide susceptibility: a study from the southern Western Ghats region of Kerala, India.," *Ann. GIS*, vol. 26, no. 2, pp. 113–131, 2020, doi: 10.1080/19475683.2020.1758207.

[12] S. Ramasamy, S. Gunasekaran, J. Saravanavel, *et al.*, "Geomorphology and landslide proneness of Kerala, India: a geospatial study," *Landslides*, 2020, doi: 10.1007/s10346-020-01562-9.

[13] N.J. Schneevoigt, S. Van Der Linden, H.-P. Thamm, and L. Schrott, "Detecting Alpine landforms from remotely sensed imagery. A pilot study in the Bavarian Alps," *Geomorphology*, vol. 93, pp. 104–119, 2007, doi: 10.1016/j.geomorph.2006.12.034.

[14] T.R. Martha, R. Priyom, K. Kirti, *et al.*, "Landslides mapped using satellite data in the Western Ghats of India after excess rainfall during August 2018," *Curr. Sci.*, vol. 117(5), pp. 804–812.

[15] L. Ayalew and H. Yamagishi, "The application of GIS-based logistic regression for landslide susceptibility mapping in the Kakuda-Yahiko Mountains, Central Japan," *Geomorphology*, vol. 65, pp. 15–31, 2004, doi: 10.1016/j.geomorph.2004.06.010.

[16] K. Khosravi, B.T. Pham, K. Chapi, *et al.*, "A comparative assessment of decision trees algorithms for flash flood susceptibility modeling at Haraz Watershed, Northern Iran," *Sci. Total Environ.*, vol. 627, pp. 744–755, 2018, doi: 10.1016/j.scitotenv.2018.01.266.

[17] Y. Song, R. Niu, S. Xu, *et al.*, "Landslide susceptibility mapping based on weighted gradient boosting decision tree in Wanzhou section of the Three Gorges Reservoir Area (China)," *mdpi.com*, vol. 8, no. 4, p. 4, 2019, doi: 10.3390/ijgi8010004.

[18] B. Guru, R. Veerappan, and F. Sangma, "Comparison of probabilistic and expert-based models in landslide susceptibility zonation mapping in part of Nilgiri District, Tamil Nadu, India," *Spatial Inf. Res.*, vol. 25, pp. 757–768, 2017.

[19] K.M.R. Hunt and A. Menon, "The 2018 Kerala floods: a climate change perspective," *Clim. Dyn.*, vol. 54, no. 3–4, pp. 2433–2446, 2020, doi: 10.1007/s00382-020-05123-7.

[20] E. Thennavan, G.P. Ganapathy, S.S. Chandra Sekaran, and A.S. Rajawat, "Use of GIS in assessing building vulnerability for landslide hazard in The Nilgiris, Western Ghats, India," *Nat. Hazards*, vol. 82, no. 2, pp. 1031–1050, 2016, doi: 10.1007/S11069-016-2232-1.

[21] K.S. Sajinkumar, S. Anbazhagan, A.P. Pradeepkumar, and V.R. Rani, "Weathering and landslide occurrences in parts of Western Ghats, Kerala," *J. Geol. Soc. India*, vol. 78, no. 3, pp. 249–257, 2011, doi: 10.1007/s12594-011-0089-1.

[22] V. Mishra and Harsh L. Shah., "Hydroclimatological perspective of the Kerala flood of 2018,." *J. Geol. Soc. India*, vol. 92, pp. 645–650, 2018.

[23] G. Bhargavi, and J. Arunnehru, "Identification of landslide vulnerability zones and triggering factors using deep neural networks – an experimental analysis," . *Advances in Computing and Data Sciences: 6th International Conference, ICACDS*, Kurnool, India, April 22–23, 2022, Revised Selected Papers, Part I. Cham: Springer International Publishing, 2022.

[24] S. Rahamana, S. Arychmay, and R. Jegankumar, "Geospatial approach on landslide hazard zonation mapping using multicriteria decision analysis: a study on Coonoor and Ooty, part of Kallar watershed, The Nilgiris, Tamil Nadu," *Int. Arch. Photogramm. Remote Sens. Spatial Inf. Sci.*, vol. XL-8, pp. 1417–1422, 2014, doi: 10.5194/isprsarchives-XL-8-1417-2014.

[25] F. Guzzetti, P. Reichenbach, M. Cardinali, M. Galli, and F. Ardizzone, "Probabilistic landslide hazard assessment at the basin scale," *Geomorphology*, vol. 72, no. 1–4, pp. 272–299, 2005, doi: 10.1016/j.geomorph.2005.06.002.

[26] L. Felipe, P. Sarmiento, M.G. Trujillo-Vela, and A.C. Santos, "Linear discriminant analysis to describe the relationship between rainfall and

landslides in Bogotá, Colombia," *Springer*, vol. 13, no. 4, pp. 671–681, 2016, doi: 10.1007/s10346-015-0593-2.

[27] P. Biju Abraham and E. Shaji, "Landslide hazard zonation in and around Thodupuzha-Idukki-Munnar road, Idukki district, Kerala: a geospatial approach," *J. Geol. Soc. India*, vol. 82, no. 6, pp. 649–656, 2013, doi: 10.1007/s12594-013-0203-7.

[28] S.S. Chandrasekaran, R. Sayed Owaise, S. Ashwin, R.M. Jain, S. Prasanth, and R.B. Venugopalan, "Investigation on infrastructural damages by rainfall-induced landslides during November 2009 in Nilgiris, India," *Nat. Hazards 2012*, vol. 65, no. 3, pp. 1535–1557, 2012, doi: 10.1007/S11069-012-0432-X.

[29] Q. He, H. Shahabi, A. Shirzadi, *et al.*, "Landslide spatial modelling using novel bivariate statistical based Naïve Bayes, RBF classifier, and RBF network machine learning algorithms," *Sci. Total Environ.*, vol. 663, pp. 1–15, 2019, doi: 10.1016/J.SCITOTENV.2019.01.329.

[30] P. Canuti, N. Casagli, L. Ermini, R. Fanti, and P. Farina, "Landslide activity as a geoindicator in Italy: significance and new perspectives from remote sensing," *Environ. Geol.*, vol. 45, pp. 907–909, 2004, doi: 10.1007/s00254-003-0952-5.

[31] W. Chen, Y. Chen, P. Tsangaratos, *et al.*, "Combining evolutionary algorithms and machine learning models in landslide susceptibility assessments," *Remote Sens.*, vol. 12, p. 3854, 2020.

[32] F. Huang, J. Zhang, C. Zhou, Y. Wang, J. Huang, and L. Zhu, "A deep learning algorithm using a fully connected sparse autoencoder neural network for landslide susceptibility prediction," *Landslides*, vol. 17, no. 1, pp. 217–229, 2020, doi: 10.1007/s10346-019-01274-9.

[33] B. Pham, D. Bui, I. Prakash, and M.B. Dholakia, "Hybrid integration of multilayer perceptron neural networks and machine learning ensembles for landslide susceptibility assessment at Himalayan area (India) using," *CATENA*, vol. 149, pp. 52–63, 2016.

[34] S.A. Rahamana, S. Aruchamy, and R. Jegankumar, "Prioritization of sub watershed based on morphometric characteristics using fuzzy analytical hierarchy process and geographical information system – a study of Kallar Watershed, Tamil Nadu," *Aquat. Proc.*, vol. 4, pp. 1322–1330, 2015.

[35] S.S. Chandrasekaran, S. Elayaraja, and S. Renugadevi, "Damages to transport facilities by rainfall induced landslides during November 2009 in Nilgiris, India," *Landslide Sci. Pract. Risk Assessment, Manag. Mitig.*, vol. 6, pp. 171–176, 2013, doi: 10.1007/978-3-642-31319-6_24.

[36] C.J. van Westen, S. Ghosh, P. Jaiswal, T.R. Martha, and S.L. Kuriakose, "From landslide inventories to landslide risk assessment; an attempt to support methodological development in India," *Landslide Sci. Pract. Landslide Invent. Susceptibility Hazard Zo.*, vol. 1, pp. 3–20, 2013, doi: 10.1007/978-3-642-31325-7_1.

Chapter 10

Machine learning paradigm for predicting reservoir property: an exploratory analysis

Saikia Pallabi[1], Deepankar Nankani[2] and Rashmi Dutta Baruah[2]

Reservoir characterization is a process of understanding different petrophysical properties of a reservoir to decide optimal location to drill a production well. Petrophysical properties means petroleum indicating subsurface characteristics that geoscientists strive to understand mainly from remotely sensed geophysical data, seismic and well logs. Machine learning (ML) had a great contribution in this field. However, the success rate varies across the reservoirs due to its characteristics, which varies from simple to sometimes very complex. The success of ML models could be enhanced with proper analysis on acquired data and modeling. In this work, we performed a case study on predicting a petrophysical property, porosity, from the analysis of geophysical data (seismic and well logs) following through multiple steps of preprocessing and modeling techniques. The methodological approach performed uses different concepts of ML, data analytic, and signal processing and achieved the desired goal to characterise the reservoir in terms of porosity. Through the case study, we investigated in depth the reservoir characterization problem from data science perspective, and we highlighted few potential research directions incorporating few challenges and opportunities that may invigorate future research of ML modeling in this field.

10.1 Introduction

Increasing demand of fossil fuels makes reservoir characterization (RC) [1] more and more crucial and at the same time challenging. Reservoir here means petroleum reservoir and characterizing it as the process of modeling the reservoir to understand the distribution of different petrophysical properties to decide a location to drill a production well that can enhance the production of an oil field. ML models [2–4] had a great contribution in modeling reservoir with geophysical data including seismic and well logs. However, the success rate varies depending on the characteristics of

[1]Rajiv Gandhi Institute of Petroleum Technology, India
[2]Indian Institute of Technology Guwahati, India

the reservoirs [5]. Seismic data is acquired through a seismic survey, which is recorded with respect to time. Well logs represent some of the important reservoir characteristics, acquired using well logging, which is recorded with respect to depth. Petrophysical properties are the reservoir characteristics that can indicate the presence of petroleum or oil. The important petrophysical properties of reservoir include porosity, permeability, fluid content, grain composition, and structure of the subsurface rocks. When seismic and well logs data of an oil field are properly processed and correlated, a good understanding of the reservoir could be obtained and hence can help in deciding the optimal position to place a production well in the field. One of the very initial works of reservoir characterization with the data-driven integration of seismic and well log was proposed by Schultz in 1994 [6]. After that, many works have been reported in this field that integrated these data sources with different ML [6–14] approaches for the prediction of porosity [15–21], permeability [15,18], water saturation [19,21,22], and many more other characteristics [18,19,21]. Although petrophysical modeling for reservoir characterisation has been widely explored till date, it still remains a challenging problem [23] due to the varying characteristics across reservoirs that include subsurface heterogeneity, spatial variability, and the presence of different complex geological features like fractures and faults in reservoirs. Hence, geoscientists still rely on other sources and disciplines [5,24] of science and engineering like Geology, Geophysics, Petrophysics, Geochemistry, Reservoir Engineering, etc. for a more certain conclusion of a reservoir. Remarkable research efforts on effectively applying of ML modeling in this field have been made in past three decades [5,24]; however, ML literature [18] in RC is comparatively weak compared to other fields of ML like vision and speech and it lacks uniformity of applied models and lacks explicit consideration and analysis of data characteristics that ultimately define the success of ML model in a particular field.

In this work, we considered a case study on the prediction of a petrophysical property, porosity. Porosity [16] indicates the fraction of total void volume present within the reservoir rock. It is an intrinsic characteristic of geological rock and one of the primary factors to identify hydrocarbon prospects in a reservoir, hence, it is a suitable petrophysical parameter to characterize a reservoir. The case study is performed on a prospect survey area, in a motive of in-depth analysis of reservoir characterization problem using ML. The ML approach followed is a framework of preprocessing and modeling techniques that are based on different concepts of signal processing, ML, and data analytics. Preprocessing techniques include well tie, seismic signal reconstruction, smoothing of well log property, and outlier removal to prepare the data well for the modeling. Different state-of-the-art regression techniques are applied for modeling and a comparative analysis is performed on the considered case study. To improve the generalization of neural network modeling, deep neural network with the regularization approaches is also investigated. At the end, postprocessing is performed to effectively visualize the distribution of porosity variation over the reservoir. Moreover, every phase of the framework have been analyzed in a data-driven manner to improve the modeling of the reservoir. Through this case study, we also identified different prominent challenges and opportunities that can open up future avenues of ML research in this field.

The paper is organized as follows: Section 10.2 provides background of geo-scientific data sources for reservoir characterization, Section 10.3 elaborates the research issues and objectives of our proposed work, Section 10.4 describes the details of the case study, Section 10.5 presents the overview of the complete ML approach to solve the problem. The modeling approaches used and the experimentation results are presented in Section 10.6. Section 10.7 presents discussion and future prospects. Section 10.8 presents a conclusion.

10.2   Geo-scientific data sources for reservoir characterization

Different fields of study perform investigation with different sources of data independently to get a comprehensive understanding of the reservoir. The fields of study include the following: geophysics, petrophysics, sedimentology, outcrop understanding, geochemistry, etc. that acquire geo-scientific data to get proper understanding of the reservoir. Among different sources, seismic survey and well logging are most common for understanding the petrophysical properties of reservoir.

10.2.1   Seismic survey

Seismic survey is one of the most popular geophysical surveys to sense the earth crust and is the primary exploration tool used by geologists as it can provide detailed 3D picture of the earth subsurface structure for the exploration of oil. It scans the earth crust with artificially generated seismic signals that are reflected by different layers of earth crust as illustrated in Figure 10.1. The survey can be performed in the ground or sea by sending shock waves to the earth crust with the help

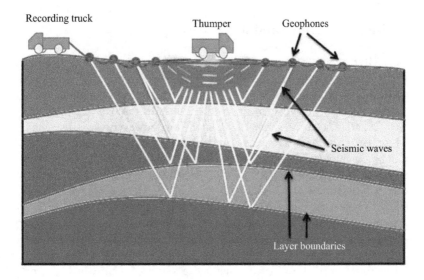

Figure 10.1   Seismic survey

of seismic wave generating sources like thumper. The produced acoustic signals flow through the earth crust and gets reflected back by different rock boundaries at different speeds. Geophones are placed on the earth surface that detects and records the reflected seismic signals from the earth to illustrate the subsurface structure. As different rocks transmit the signals differently, the measurement of the transmitted signals can reveal different properties and location of each rock layer. With the advancement of sensor technology, it is possible to sense and scan the earth crust with apparently good resolution seismic data and hence it is possible to explore even the reservoirs present deep beneath the earth crust.

10.2.2   Well logging

Logging is the process of continuous recording of the variations of the physical properties of a borehole in a depth wise manner. Log records can be of two types, namely geological logs and geophysical logs. Geological logs are the records that are drawn from the cutting rock samples that are brought to the surface which is called core sampling and geophysical logs are the recordings performed by lowering a set of probes containing wireline logging tools into the well which is called well logging. Well logging procedure attaches the probe with different wireline logging tools to record the variation of physical properties throughout the depth of the well. Different electrical, sonic, or nuclear logging tools can be used to estimate the interested physical properties of the well that includes density, pressure, fluid content, porosity, etc. Well logging is a widely used data acquisition technique that can give the quantitative measurements of the petro-physical properties throughout the borehole. Recorded data is of high resolution but the coverage is small limited to the borehole location. The technique is costly and not possible to apply to the entire area of survey.

10.3   Research issues and objectives

Different ML techniques [25] are applied in the literature for the prediction of petro-physical properties that include decision tree [26], random forest [27], support vector regression [28], and neural networks [26]. Recently, the soft computing have also been advanced in this field to enhance the ability to discover and estimate new reserves [29,30]. The success rate of ML modeling, however, varies over the reservoirs [26,31,32] and there is no specific technique that will always yield the best result. The outcome of the modeling technique subject to the kind of data used. According to a GeoExpro Magazine [33] in 2015, the global success rate of oil exploration is disheartening. More than 50% of the wells drilled were unable to produce hydrocarbons. Over the years, the success rate is even decreasing due to moving towards the reservoirs deep down the earth crust having complex compositions. Drilling a well in the earth crust to extract hydrocarbon costs around hundreds of crores of rupees. Hence, drilling a dry well can incur huge loss to the industry. It is crucial to focus on how to increase the success rate of well, which can only be possible with effective understanding and modeling of oil reserves and reservoir characteristics. Some of the challenges that are posed while developing ML for predicting reservoir characteristics from seismic and well logs are listed below:

• Reservoir characteristics that are recorded in the form of well logs include porosity, permeability, water saturation, grain and sand fractions and share a complex and nonlinear relationship with the seismic data. For modeling a complex and nonlinear relation, ML models demand a huge set of data samples. In newly developed oil fields, collecting such data is costly and also time consuming. This type of scenarios limits the ability of ML methods to effectively model the reservoir.

• Due to the heterogeneous nature of earth subsurface, the characteristics of reservoirs vary away from the well locations [34]. Predicting the petrophysical properties of a well (test well) that is farther from the training well locations can be difficult if the test well location is not in their neighbourhood. The test well is called as blind well as the trained model has not seen any data characteristic of it during its training. Hence, the blind well prediction can be challenging if the blind well distribution largely varies from the trained wells distribution.

Our work focuses on the estimation of petro-physical properties to effectively characterize a given reservoir using seismic data and well logs. The objective of this research is to address the afore-mentioned challenges by investigating and developing different ML paradigm for the estimation of petro-physical properties. We provided an initial analysis of basic ML models of the regression for the prediction of petro-physical properties in the considered oil field to analyze how effectively they can predict on our considered dataset, and on which scenarios we can expect a good performance. To improve the performance of modeling, we effectively pre-processed the dataset using signal processing approaches. We compared different standard preprocessing approaches for preparing the seismic and well logs for modeling. Preprocessing techniques are applied basically to calibrate the data sources that are different in characteristics. Through the case study of ML models on petrophysical property modeling, we investigated the reservoir characterization problem from data science perspective and highlighted potential research directions incorporating few challenges and opportunities.

10.4   Description of the case study

For the case study, we considered a survey area of an oil field provided by Oil and Natural Gas Corporation Limited (ONGC), India, the base-map of which is shown in Figure 10.2. The survey data constitutes of 3D seismic and well log data available over the seven well locations as denoted by the seven spatial points. The available wells (Well 1, Well 2, Well 3, Well 4, Well 5, Well 6, and Well 7) contain the information of porosity variation across the well depth in terms of well logs. The schematic diagram with differences in seismic and well log signal is provided in Figure 10.3.

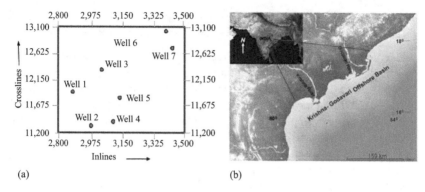

(a)   (b)

Figure 10.2   Geological setting of survey area [35]. (a) Survey Area, (b) Location of Krishna Godavari Basin.

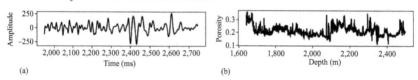

(a)   (b)

Figure 10.3   Representative plots of seismic and well log; seismic: timewise measurement, low sampling rate, contains low-frequency, low resolution, large area coverage; well log: depthwise measurement, high sampling rate, contains high-frequency, high resolution, small area coverage. (a) Seismic Signal recording w.r.t. time, (b) Well log property (Porosity) recording w.r.t. depth.

10.4.1   Geological background of the survey area

The base-map of the survey area is provided in Figure 10.2a, which is located in the region of Krishna–Godavari basin oil field, having seven exploratory and production wells. The Krishna–Godavari Basin is located in the South-Eastern part of India. The basin lies across the Bay of Bengal and the blue rectangle in the figure is the approximate location of the Basin, east of Hyderabad [35]. The sediment contents of the basin are of thick sequences formed with several cycles of deposition. The basin has several sub-basins located in on-land as well as in offshore and has the potential reservoirs of oil and gas. To the southeast, the basin extends into the deep water of the Bay of Bengal. The area is drained by two major rivers: Krishna and Godavari. A large stratigraphic section of the basin from oldest Permo-Triassic Mandapeta Sandstone in on-land to the youngest Pleistocene channel levee complexes in deep water offshore has the great potential to host large volumes of hydrocarbon reserves [36].

10.5   ML for reservoir characterization: the proposed approach

The proposed workflow for the prediction of the reservoir property is provided in Figure 10.4. One of the challenging tasks for ML modelling is to calibrate the

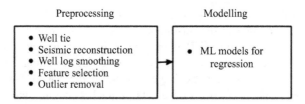

Figure 10.4   Reservoir characterization workflow

seismic and well log due to the differences in characteristics as illustrated in Figure 10.3. Proper processing is required to integrate them to derive useful information about the reservoir. Preprocessing is performed to calibrate the data and bring them both to similar characteristics. Relation between seismic attributes and porosity is approximated with different ML models of regression analysis.

For proper ML modeling, it is very crucial to analyze the data and prepare it accordingly. Analysis and processing must go hand in hand to prepare the data suitable for modeling. Different state-of-the-art signal processing and statistical techniques are applied for performing the defined steps of preprocessing (as in Figure 10.4), and their behavior on our dataset have been inspected accordingly.

10.5.1   Well tie

Seismic and well data recordings are performed in different domains, time and depth respectively. Hence, calibration of these data sources is possible by bringing them to the same domain. The well data which is in-depth is converted to time using well-to-seismic tie [1]. It is performed by generating synthetic seismic trace corresponding to a well location and then matching it with the real seismic trace. The best match of the synthetic with the real one provide the depth to time relation. The synthetic trace is generated from the basic well log signals sonic log P-wave (V_p) and bulk density (ρ), which are available over all the wells. These signals together provide the information about the acoustic impedance that provide the reflection coefficient series ($R_0(t)$) of the subsurface, which when convolved ($*$) with an estimated source wavelet ($w(t)$) provide the synthetic seismic trace $T(t)$. The mathematical representation of the generation of synthetic seismic trace is provided in (10.1):

$$T(t) = R_0(t) * w(t) + n(t) \tag{10.1}$$

where $n(t)$ is the noise component. Once the depth and time relation is known, we proceed with integrating the seismic and well data to understand the underlying relationship between them.

10.5.2   Seismic signal reconstruction

The sampling rate of seismic is low compared to well logs. However, it is essential to have same sampling interval for both the data sources for modelling the relation. Hence, we up-sampled the seismic signal to the level of well log using seismic signal reconstruction, so that we need not compromise with the samples of the wells. The reconstruction of the seismic signal is performed by cubic spline interpolation. Spline interpolation [37] is applied as the seismic signal has the characteristics of local and abrupt changes. Using higher order polynomial to interpolate the data may oscillate largely between the data points that it may increase the reconstruction error of the data points between the interval. Whereas cubic spline interpolation fit a series of unique cubic polynomial functions between each data point to obtain a continuous and smooth curve that can be helpful for approximating the actual graph of the seismic signal. Figure 10.5 provides the graph of the seismic traces corresponding to respective well before and after interpolation. The dots represent the original values, and the curves represent interpolated seismic traces using cubic spline interpolation. We can observe that cubic spline interpolation smoothly approximating the seismic signal, which very closely satisfies the smoothness characteristics of seismic signal, along with preserving the local and abrupt changes.

10.5.3   Smoothing of well log

Seismic attributes are smooth signals compared to the well logs. Well log data is of high frequency compared to seismic signals. It is difficult to predict high frequency signal from the low frequency one according to the information theory [38]. Hence, we applied smoothing on the well log using signal processing concepts of filtering using mean, median, mean–median, and Fourier regularization [39–42]. The comparative results are shown in Figure 10.6. From the comparative analysis of the

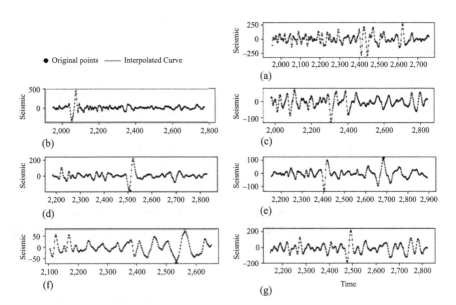

Figure 10.5   Interpolated seismic traces on the original sample points for Well 1 to Well 7 (from a to g)

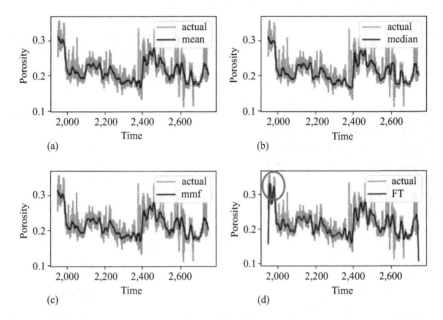

Figure 10.6   Smoothing porosity log using different filtering techniques. (a) Mean filtering, (b) median filtering, (c) mean median filtering, and (d) fourier regularised.

filtered logs, Fourier regularization seems to follow better trend of the original trace compared to the other filtering approaches, i.e., mean filter, median filter, and mean–median filter. Fourier transform could more smoothly approximate, even the abrupt peaks of the signal (as shown by the red circle), and from the domain of geoscience, these changes have some inevitable information that needs to be considered. Fourier regularization is based on frequency selective filtering technique performed by analyzing a signal in the frequency domain. With this technique, we selectively remove the frequency components from the signal and hence it helped to match the frequency contents of both seismic and well log signals. Also, removal of high frequency components from a signal, sometimes help to remove high frequency noise. When we analyzed the seismic and well log from all the available wells, the seismic signal contains around lower 10% of the total frequency components as in well log. After removing the rest high frequency components using Fourier regularization, the regularized plot is shown in Figure 10.6(d), that is smoothly approximating the variation of the original porosity log. However, the Fourier regularization imposed some ill samples in the endpoints of the well log. Hence, we remove those samples to maintain the quality of the data.

10.5.4   Seismic attributes selection

Seismic signal itself does not carry sufficient information about the underlying subsurface properties as it does not see the earth subsurface with the same resolution as the well log sees. Hence, we considered few analytically derived seismic attributes [7,43]: amplitude envelope (A1), amplitude-weighted cosine phase (A2), amplitude-weighted frequency (A3), amplitude-weighted phase (A4), derivative (A5), dominant frequency (A6), instantaneous frequency (A7), instantaneous phase (A8), integrate (A9), integrated absolute amplitude (A10), quadrature trace (A12), and seismic amplitude (A13), along with two more attributes generated by seismic inversion [21], P-impedance (A11) and VpVs (A14). All the attributes derived here not necessarily contain relevant information about the porosity prediction. To understand the importance of these derived attributes about the porosity, we used different statistical methods [44,45] of feature importance that includes Pearson correlation (P_Corr), F Test (F_Test), and Mutual Information (M_I) and the comparison results are provided in Figure 10.7 with its normalised values. P_Corr and F_Test could capture only linear dependency whereas M_I [46] can capture any kind of dependencies including nonlinear too. It can be observed from Figure 10.7 that all the methods marked that attributes A6, A11, and A14 captured the dependencies better than the other attributes including main seismic signal (A13) about the porosity. Better the dependency, the better can be the porosity prediction using these attributes. These higher dependency attributes together seems to carry high information compared to other attributes. Hence, computationally efficient model can be build using these selected important features. Another important consideration while selecting the attributes is that the feature analysis methods as listed above do not provide information about multicollinearity. Having multicolinear attributes can affect the performance of training models. So, we must remove the multicollinearity of features well before training the models for data. The colinearity of the attributes can be identified by calculating the correlation coefficients (CC) between the attributes. The CC for the attributes comprising the input and target is provided in Figure 10.8. Larger the box in the corresponding location higher the correlation. White box means positively correlated and darker box means negatively correlated. It is observed that the amplitude-weighted cosine phase (A2) and seismic (A13) is highly correlated with each other. Consideration of these kinds of attributes together in prediction should be avoided. Considering the above two aspects of feature importance and collinearity, we considered A6 (dominant frequency), A11 (P-impedance), and A14 (VpVs) as the final selected features for the ML modeling.

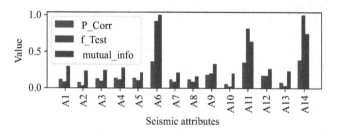

Figure 10.7   Seismic feature importance of porosity

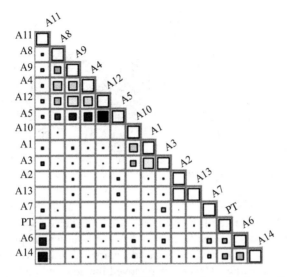

Figure 10.8   Correlation coefficients (CC) among the seismic attributes and target

10.5.5   Outlier removal

An outlier is a data point that differs in the characteristics from normal samples of a population. An outlier can be observed due to experimental or measurement errors, high noise contamination, etc. In multivariate data, outlier detection can be difficult with an increase in the number of dimensions and the level of data contamination [47]. However, detection and removal of it is necessary for ML models' performance as many estimators are largely sensitive to the presence of outliers. The outlier detection is commonly performed with distance measures [48] from the mean of the data samples. The results using the commonly used distance measures, Euclidean distance (ED), and the Mahalanobis distance (MD) [49,50] are provided in Figure 10.9. The samples are plotted on the first two components obtained using principal component analysis (PCA). Although we could not provide the plot of whole data samples, however, we tried to show few samples to show the characteristic difference obtained by both the techniques. In multivariate data, covariance consideration can be important as can be observed from Figure 10.9. As, ED does not consider the covariance present between the variables, so it may incorrectly detect outliers as compared to MD, which takes consideration of the covariance [51] among the attributes. If a point in a space is represented by X, and the mean of the distribution from where X belongs is μ, then the MD of point X from mean is calculated by (2):

$$MD(X) = \sqrt{(X - \mu)\sum^{-1}(X - \mu)^T}$$

(10.2)

The outliers detected are presented in Figure 10.9 with red samples. In Figure 10.9(b), ED detected samples with the consideration of Euclidian distance,

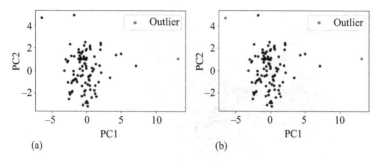

Figure 10.9   Outlier removal. (a) MD and (b) ED.

so it may consider some valid samples as outlier if distance measure is low and in case of high distance measure, it can even leave the outlier samples as clean samples. Comparatively, MD provided better detection of outlier as it considers covariance to declare a outlier. Hence, we removed the detected outliers using MD to keep our data clean.

10.6   Experimental results and analysis

10.6.1   Statistical data analysis

We have performed statistical analysis to gain understanding of the data characteristics and to determine the modeling technique. Keeping in consideration that our data is distributed across different spatial locations (seven wells), we have analysed the descriptive statistics of the selected seismic attributes (dominant frequency, VpVs, and P-impedance) and the output property, porosity, across different wells considered. The descriptive statistics, also called as univariate statistical analysis, are presented in Table 10.1 and the corresponding line curve are presented in Figure 10.10. The table contains the description of the attributes in terms of number of samples (count), statistical mean (mean), standard deviation (std), minimum value of the attributes (min), maximum value of the attributes (max), corresponding to the wells considered. However, line curves are provided for mean, std, min, max to show the variations of these attributes across the wells. The description can be helpful to understand the attributes characteristics across different wells. It is apparent from the results that the attributes do not follow similar statistical characteristics across the wells. If this is the case, normalization of attributes using these parameters may not be suitable for the real-world data, especially min–max parameters. This is because the min–max values of the test data may go out of range from the normalized parameters obtained from the observed data. This demands careful selection of normalization technique. For understanding the data distribution of different attributes, kernel density estimation plots are provided as shown in Figure 10.11. It is apparent that the data distribution of the wells are different across the wells. This is very critical in terms of model

Table 10.1   Univariate statistical analysis of attributes well-wise

	Well 1	Well 2	Well 3	Well 4	Well 5	Well 6	Well 7
			Dominant frequency (DF)				
Count	5,645	6,000	6,455	4,730	5,605	4,089	4,983
Mean	30.86	25.43	19.00	24.95	21.97	22.62	27.18
Std	7.487	5.7586	5.104	5.467	7.120	5.873	5.392
Min	19.41	18.28	11.51	17.73	12.69	12.72	16.49
Max	51.17	39.48	33.33	34.15	34.87	32.78	35.89
			P-Impedance (P-Imp)				
Count	5,645	6,000	6,455	4,730	5,605	4,089	4,983
Mean	4,716.55	4,766.67	5,063.86	5,063.51	5,052.07	5,192.58	5,102.64
Std	546.99	555.27	570.13	376.01	475.39	316.71	513.93
Min	3,617.27	3,086.12	4,179.50	4,044.74	4,187.00	4,653.01	4,216.43
Max	6,132.09	6,173.02	6,023.05	5,863.10	5,995.20	5,789.97	6,523.75
			VpVs				
Count	5,645	6,000	6,455	4,730	5,605	4,089	4,983
Mean	2.452	2.615	2.3382	2.426	2.374	2.506	2.381
Std	0.2474	0.2472	0.2635	0.1911	0.2293	0.1335	0.1848
Min	1.8435	1.8905	1.93660	1.8185	1.8008	2.2360	1.7867
Max	3.0097	3.1652	2.8573	2.7168	2.7963	2.7408	2.6733
			Porosity (PT)				
Count	5,645	6,000	6,455	4,730	5,605	4,089	4,983
Mean	0.21364	0.24346	0.19153	0.2422	0.2241	0.2623	0.2673
Std	0.03338	0.02633	0.04454	0.0307	0.0339	0.0278	0.0335
Min	0.1125	0.1032	0.06008	0.1537	0.1297	0.1831	0.1106
Max	0.3456	0.3070	0.2966	0.3264	0.3344	0.3539	0.3349

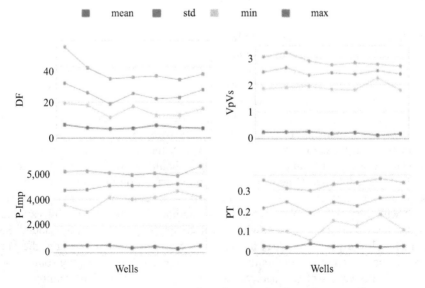

Figure 10.10   Univariate statistical analysis of attributes well-wise

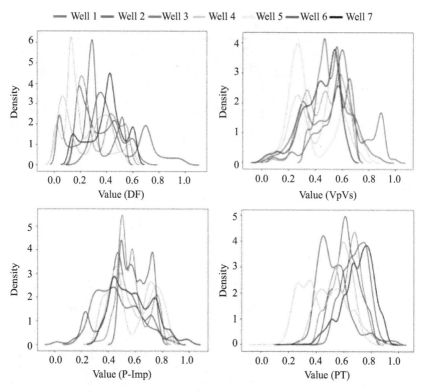

Figure 10.11   Kernel density estimate of different attributes

building. Generally, predictive models expect the training and test data to come from similar distribution. On combining all the wells, the data distribution of the attributes looks like in Figure 10.12. The attributes behave like following the normal distribution, with the lack of enough data samples in some regions. The distribution can be more complete with the increase number of wells and the samples. Hence, considering the distribution of input attributes, we normalized the attributes with z-score normalization, which corresponds with the normalization parameter mean (μ) and standard deviation (σ). However, the target attribute, porosity, is normalized within the range of 0–1, so that it should be suitable for the modeling methods including neural network with sigmoid activation function, that demands target in the range of 0–1. The considered field contains 20–40% porosity variations with very less samples going out of this bound. Hence the output variables mostly lies between 0.2 and 0.4, with very less samples going out of this bound. So, to make all the possible values to normalize between 0 and 1, we kept the normalized parameters with minimum 0.1 and maximum 0.9 with keeping some offset to handle the values of test data which comes out of range of the minimum and maximum values of the train data.

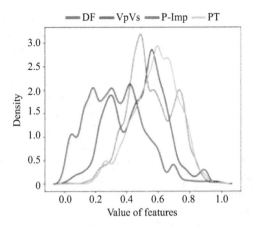

Figure 10.12   Kernel density estimation for attributes with all wells combined

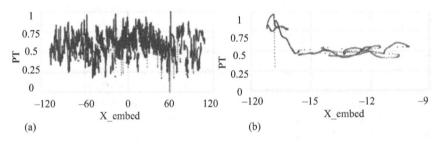

(a)   (b)

Figure 10.13   Porosity (PT) plot in reduced dimension X_{embed}. (a) All samples considered. (b) First 1,000 samples.

We have also analyze the patterns of the data to prepare modeling hypothesis. With a motive to visualize the behavior of the data we employed t-SNE (T-distributed stochastic neighbor embedding) [52] plot as provided in Figure 10.13. It is the representation of target property in reduced dimension denoted as xembed. Two points closer on the *X*-axis indicates that the two samples are closer to each other in the feature hyperspace. Figure 10.13(b) is of first 1,000 samples of Figure 10.13a, which is the t-SNE plot of all the samples considered. This is done to zoom in the samples to clearly understand how the patterns vary according to its neighboring samples. it is observed that changing of porosity is of smooth changing curve but with very complex pattern. This suggests that there exists a complex nonlinear relationship between input attributes and the target property. Nonlinear ML model capable to handle nonlinear complexity seems to be useful with this data.

10.6.2   Results and analysis of ML modeling

In order to understand the relationship between seismic attributes (input) and porosity (output), different ML models for regression data are explored and compared subsequently. The training dataset for ML models is obtained with 80% of the total samples from the combined data from all the wells. It was generated by shuffling the data first, to remove any possible trend along the depth, rest 20% is considered for testing. The most cumbersome task with machine learning models is to deal with fiddling with the models, tuning their parameters. So, from the training data, around 20% data are taken out for validation set to tune the hyperparameters. The generalization capability of the trained model is evaluated during testing phase. The model performances are evaluated in terms of root mean square error (RMSE), coefficient of determination (R-square), and correlation coefficient (CC) measures. Several runs of training, testing, and validation phases have been carried out in order to decide the best set of hyperparameters and parameters for every ML model. The input attributes for ML models constitutes the selected features as mentioned in Section 10.5.4, and the main seismic signal. Different ML regression models [18,26,53–57] are explored on our dataset. We have considered artificial neural network (ANN), k-nearest neighbors regressor (KNN), support vector regression (SVR), decision tree (DT), and linear models from Scikit Learn Library [58]. Even though it is clear from the statistical analysis of Section 10.6.1 that the underlying relationship among data is nonlinear, we have also explored linear models as in some literature linear models have also been effectively used for the prediction of petrophysical properties [18], and linear model can provide the information about the percentage of linear information the data contains. Results of the linear methods that include stochastic gradient descent (SGD) regressor, Ridge, Lasso, and Elastic Net are provided in Table 10.2 with their comparisons.

All the linear models provided almost comparable performance. It is apparent from the result that the data is having very less fraction of linear dependency between input and output (around 22%), as visible from the R-square score. However, R-square score in nonlinear model is not a good practice as it sometimes draw false conclusions about nonlinear model performance [59]. Hence, we discard this metric for nonlinear models evaluations. Elastic Net [60], which has the capability to outperform other linear models in large dataset (as it combines the advantages of both ridge and lasso regression that take care of both L1 and L2 regularization, respectively), is considered to present the prediction curve as provided in Figure 10.14a. The non-linearity in the data has not been captured by the model as can be seen in predicted vs. actual plot. The patterns that could not be

Table 10.2   Comparison of linear regression models

Linear models	RMSE	R-square	CC
Linear regression	0.138	0.22	0.47
Huber regression	0.138	0.22	0.47
SGD regressor	0.138	0.22	0.47
Ridge	0.138	0.22	0.47
Lasso	0.139	0.21	0.47
ElasticNet	0.138	0.22	0.47

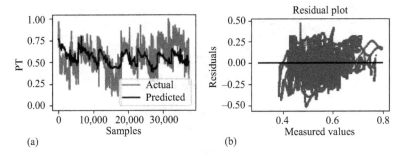

Figure 10.14   Analysis plots of linear model (ElasticNet). (a) Prediction plot and (b) Residual plot.

Table 10.3   Comparison of nonlinear regression models

Non-linear models	RMSE	R-square	CC
ANN	0.077	0.754	0.87
KNN (uniform)	0.0035	0.99	0.99
KNN (distance)	0.003	0.99	0.99
SVR	0.032	0.95	0.97
DT	0.0173	0.987	0.99

understood by the linear models are provided in Figure 10.14(b), which is the error plot corresponding to the fitted values. It can be observed that the residual error variance is not constant over the measured values. This indicates Heteroskedasticity [61] issue that indicates the presence of nonlinearity in the data which has not been captured by the model. Hence, nonlinearity modeling is essential with this kind of data as the linear models could not explore the input–output relation properly. Experiments are performed with the popular nonlinear regression models, and the results are presented in Table 10.3. The k-nearest neighbors regression surprisingly outperformed other algorithms, SVR, DT, and ANN models. The model achieved best result with RMSE of 0.0035 and CC of 0.99 on the test data, even in very little time (of 1.2 s) compared to other models. Two variations of KNN based on the weighting of its neighbor samples are applied, one with uniformly weighting the considered k-neighbours and the other with weighting the neighbours based on its distance from the considered data point. Distance-based weighting of neighbors is comparatively better approach with our dataset as can be observed from Table 10.3.

When compared both the approaches with the variation of its hyperparameter, i.e., number of neighbours, it is observed that the distance-based weighing provided better results in every hyper parameter configurations as can be observed in Figure 10.15. This observation indicates that the prediction can get better if the test samples are from those regions of the sample space, more numbers of nearby

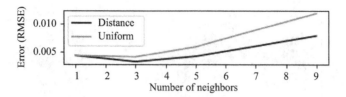

Figure 10.15 KNN performance with variation of neighbors

Table 10.4 *Comparison of regression models*
for blind well prediction

ML models	RMSE	CC
Linear	0.055	0.42
ANN	**0.05**	**0.50**
KNN	0.052	0.34
SVR	0.056	0.004
DT	0.064	0.152

Note: The best performance metrics values are highlighted
with bold fonts.

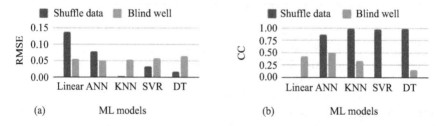

Figure 10.16 Comparison of models on shuffled and blind well testing

training patterns are available for the model. This motivates us to investigate for the prediction results for blind location/well, where the data is expected to comes from a location further from the wells considered during training. Due to geological characteristics heterogeneity, it is assumed that the characteristics of data can vary from one location to another, hence the prediction can be difficult in this scenario.

Separating Well 1 samples from all the other wells, we define Well 1 as blind well as its samples are not considered in training dataset. The result of blind well prediction is provided in Table 10.4, and its comparison to prediction on shuffled data (as provided in Table 10.3) is provided in Figure 10.16.

It clearly indicates that KNN that performed best could not perform that good in blind well prediction as its performance is completely based on the availability of nearby samples in its samples space, which seems is not properly present here due to the consideration of test well from different location having different characteristics. However, improvement in results (both in terms of RMSE and CC) could be seen with ANN model in blind well prediction. This can be because ANN basically tries to understand the underlying functional relationship between input and output without consideration of the concept of nearby samples. If we provide good set of samples that are sufficient to understand the underlying distribution of data, ANN can possibly outperform.

10.6.3 Performance comparison of shallow vs. DNN model

In the literature, it is popular that deep model of ANN (ANN with multiple hidden layers) provided state-of-the-art performance in approximating the complex non-linear function [62–64] in many different domain of applications. But one of the concern is that the deep ANN model is more prone to overfit the train data if sufficient measure is not taken care. Dropout [65] is one of the popular measures to cope with the over-fitting of train data. Hence, here we employed deep model of ANN to improve the generalization, with the introduction of dropout to take care of overfitting. Batch normalization [66] is also employed to scale and re-center inputs to each layer, which ultimately helps in retaining gradients and faster learning.

The representative architecture of the deep model of ANN (DNN) with the proposed regularization scheme (dropout + batch normalization) is provided in Figure 10.17(a) and results with varying depth on the blind well prediction are provided in Figure 10.17(b). DNN provided better result with increased depth, which is shown by different colors for different value of depths. We also observed in the result that comparable result can be achieved by deep neural network with less number of computational units as compared to the shallow neural network. However, the complex nature of the data that we have stopped us in improving generalization results further. We could obtained best result with 2 layer deep neural network with a RMSE of 0.049.

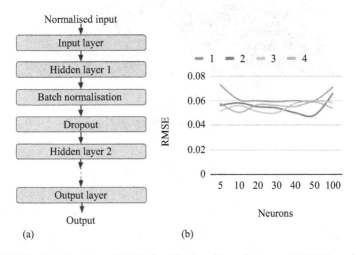

Figure 10.17   Realization of DNN for blind well prediction: (a) DNN architecture and (b) results on varying layers

The ultimate goal of the modeling is to generate the spatial distribution of the petrophysical property to visually interpret the underlying reservoir. In order to generate the spatial distribution of porosity, the best model obtained above (DNN) is considered and the porosity values are predicted for each spatial point corresponding to available seismic traces. The 2D visualization of generated porosity for Inline no. XX78 is provided in Figure 10.18. However, when we visualize seismic data as shown in Figure 10.19, we see that seismic attribute varies smoothly across the subsurface. Hence, we understand that the spatial variation of petrophysical property must be smooth, it cannot vary so abruptly as in Figure 10.18. Similarly the volumetric visualization of porosity along with other petrophysical properties can be made to detect the potential zones of hydrocarbon presence. Hence, we applied this rationale with a smoothing technique, median filtering, and the result of filtering as shown in Figure 10.20 that enhanced the continuity of the generated subsurface.

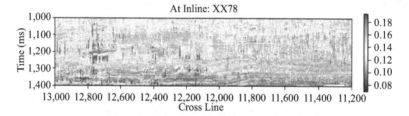

Figure 10.18   Porosity distribution for Inline XX78

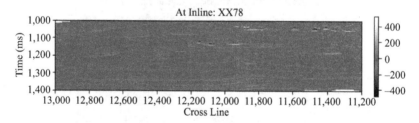

Figure 10.19   Seismic data for Inline XX78

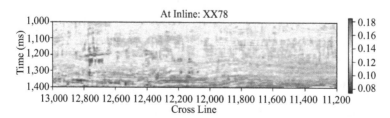

Figure 10.20   Porosity distribution for Inline XX78 (with filtering)

10.7   Discussion and future prospects

We have provided a complete ML workflow with analysis, for the prediction of porosity from seismic attributes. The overall summary of the result is provided in Table 10.5, which illustrates the improved performance of DNN in our case study. Other observations, we came across through this case study, are mentioned below:

- Deep ANN model applied here seems a promising approach in regression problem including the one with the proposed regularization scheme (dropout + batch normalization). The model has the capability of providing good generalization even in limited data. However, we must take care that with the increase in layer the model should not overfit. For the complex regression dataset, the deep model of ANN can be of good advantage as it can approximate the same function as shallow ANN in reduced number of computational units.
- Regression models are restrictive in nature as it makes assumptions of the underlying data distribution based on the data it sees. It fails to deliver good results with data sets which do not fulfill its own assumptions. One of the challenges we may face when considering the reservoir data is heterogeneity issue of the reservoir, where the characteristics of the data vary from place to place. Due to this, generalizing the regression model to the places further from the well locations can be difficult. This challenge may fall in the field of handling incomplete data [67], which is out-of-scope of these applied ML models, as the ML models expect the test data comes from the same distribution as seen by the train data. The future work may involve investigating ML techniques for incomplete or limited data scenarios.
- The prepossessing steps that include signal reconstruction, smoothing, feature extraction and selection, and outlier removal are important to enhance the learning capability of models in carrying out mapping between seismic attributes and reservoir property successfully. Moreover, the integration of data sources (seismic and well logs) using well tie plays a crucial role in the performance of the complete process. Enhancing the prepossessing (with the incorporation of domain knowledge) can bring an improvement of mapping of the data.

Table 10.5   Summary of ML models in predicting porosity in well location 1

ML models	RMSE	CC	Parameters
Linear	0.055	0.42	–
ANN	0.05	0.50	Neurons: 200, activation function: sigmoid, early_stopping: true
KNN	0.052	0.34	n_neighbors: 2, weights: "distance"
SVR	0.056	0.004	gamma: 200.0, epsilon: 0.0001
DT	0.064	0.152	max_depth: 40
DNN	**0.049**	**0.51**	2 Layers with 50 neurons, ReLU activation, dropout: 0.5, early stopping

- The spatial distribution of reservoir property enable us to understand the distribution of underlying subsurface properties from place to place. This approach can provide the details of high porosity zones. Thus, can support in identifying potential drilling locations in an oil field.

10.8 Conclusion

This study investigated a challenging and critical real-world regression problem, reservoir characterization, with an ML approach. Sensors data (seismic and well logs) are integrated in order to generate underlying porosity distribution of a considered prospect area. The experiments and analysis have been performed on the field scale data and different signal processing and machine learning concepts are introduced to overcome the different aforementioned challenges. Through this case study, we put an effort to introduce the researchers in this domain with the opportunities and challenges of using different ML models. We hope that our discussions and the major assessment techniques applied here can serve the future ML researchers in this domain as a detailed resource to solve the problem of Reservoir Characterisation in any kind of reservoir. Also, we hope that our insights on the challenges and the prospects identified in this research may guide the potential research directions in this field in particular, and data engineering and ML research in general.

Acknowledgment

This study was supported by Geodata Processing & Interpretation Centre (GEOPIC), Oil and Natural Gas Corporation (ONGC) Limited, Dehradun, India (M 984563). The authors are grateful to Mr Sanjai Kumar Singh and Mr P. K. Chaudhury of GEOPIC, ONGC for their valuable suggestions.

References

[1] Lines LR and Newrick RT. *Fundamentals of Geophysical Interpretation.* Society of Exploration Geophysicists; 2004.
[2] Mishra S and Datta-Gupta A. *Applied Statistical Modeling and Data Analytics: A Practical Guide for the Petroleum Geosciences.* Elsevier; 2017.
[3] Bhattacharya S and Mishra S. Applications of machine learning for facies and fracture prediction using Bayesian Network Theory and Random Forest: Case studies from the Appalachian basin, USA. *Journal of Petroleum Science and Engineering.* 2018;170:1005–1017.
[4] Sebtosheikh MA and Salehi A. Lithology prediction by support vector classifiers using inverted seismic attributes data and petrophysical logs as a new approach and investigation of training data set size effect on its performance in a heterogeneous carbonate reservoir. *Journal of Petroleum Science and Engineering.* 2015;134:143–149.

[5]   Aminzadeh F. Applications of AI and soft computing for challenging problems in the oil industry. *Journal of Petroleum Science and Engineering.* 2005;47(1–2):5–14.

[6]   Schultz PS, Ronen S, Hattori M, *et al.* Seismic-guided estimation of log properties (Part 3: a controlled study). *The Leading Edge.* 1994;13(7):770–776.

[7]   Hampson D, Todorov T, and Russell B. Using multi-attribute transforms to predict log properties from seismic data. *Exploration Geophysics.* 2000;31 (3):481–487.

[8]   Saggaf MM, Toksöz MN, and Mustafa HM. Estimation of reservoir properties from seismic data by smooth neural networks. *Geophysics.* 2003;68 (6):1969–1983.

[9]   Tonn R. Neural network seismic reservoir characterization in a heavy oil reservoir. *The Leading Edge.* 2002;21(3):309–312.

[10] Chaki S, Routray A, and Mohanty WK. Well-log and seismic data integration for reservoir characterization: a signal processing and machine-learning perspective. *IEEE Signal Processing Magazine.* 2018;35(2):72–81.

[11] Naeini EZ, Green S, Russell-Hughes I, *et al.* An integrated deep learning solution for petrophysics, pore pressure, and geomechanics property prediction. *The Leading Edge.* 2019;38(1):53–59.

[12] Alaudah Y, Michalowicz P, Alfarraj M, *et al.* A machine learning benchmark for facies classification. *Interpretation.* 2019;7(3):1–51.

[13] Gogoi T and Chatterjee R. Estimation of petrophysical parameters using seismic inversion and neural network modeling in Upper Assam Basin, India. *Geoscience Frontiers.* 2018;10:1113–1124.

[14] Cheng C, Yu W, and Bai X. The research on method of interlayer modeling based on seismic inversion and petrophysical facies. *Petroleum.* 2016;2(1):20–25.

[15] Iturrarán-Viveros U, and Parra JO. Artificial neural networks applied to estimate permeability, porosity and intrinsic attenuation using seismic attributes and well-log data. *Journal of Applied Geophysics.* 2014;107:45–54.

[16] Jalalalhosseini S, Ali H, and Mostafazadeh M. Predicting porosity by using seismic multi-attributes and well data and combining these available data by geostatistical methods in a South Iranian Oil Field. *Petroleum Science and Technology.* 2014;32(1):29–37.

[17] Oliveira L, Pimentel F, Peiro M, *et al.* A seismic reservoir characterization and porosity estimation workflow to support geological model update: presalt reservoir case study, *Brazil. First Break.* 2018;36(9):75–85.

[18] Saikia P, Dutta Baruah R, Singh SK, *et al.* Artificial neural networks in the domain of reservoir characterization: a review from shallow to deep models. *Computers & Geosciences.* 2019;135:104357.

[19] Saadu Y and Nwankwo C. Petrophysical evaluation and volumetric estimation within Central swamp depobelt, Niger Delta, using 3-D seismic and well logs. *Egyptian Journal of Petroleum.* 2018, 27(4), pp. 531–539.

[20] Kumar R, Das B, Chatterjee R, *et al.* A methodology of porosity estimation from inversion of post-stack seismic data. *Journal of Natural Gas Science and Engineering.* 2016;28:356–364.

[21] Gogoi T and Chatterjee R. Estimation of petrophysical parameters using seismic inversion and neural network modeling in Upper Assam Basin, India. *Geoscience Frontiers.* 2019;10(3):1113–1124.

[22] Amiri M, Ghiasi-Freez J, Golkar B, *et al.* Improving water saturation estimation in a tight shaly sandstone reservoir using artificial neural network optimized by imperialist competitive algorithm–a case study. *Journal of Petroleum Science and Engineering.* 2015;127:347–358.

[23] Jia A, He D, and Jia C. Advances and challenges of reservoir characterization: a review of the current state-of-the-art. In: *Earth Sciences.* IntechOpen; 2012.

[24] Yu X, Ma YZ, Psaila D, *et al.* Reservoir characterization and modeling: a look back to see the way forward. *Uncertainty Analysis and Reservoir Modeling, AAPG Memoir.* 2011;96:289–309.

[25] Ahmadi MA and Chen Z. Comparison of machine learning methods for estimating permeability and porosity of oil reservoirs via petro-physical logs. *Petroleum.* 2019;5(3):271–284.

[26] Tan F, Luo G, Wang D, *et al.* Evaluation of complex petroleum reservoirs based on data mining methods. *Computational Geosciences.* 2017;21(1): 151–165.

[27] Anifowose F, Labadin J, and Abdulraheem A. Ensemble learning model for petroleum reservoir characterization: a case of feed-forward back-propagation neural networks. In: *Pacific-Asia Conference on Knowledge Discovery and Data Mining.* Springer. 2013. p. 71–82.

[28] Anifowose FA, Ewenla AO, Eludiora SI, *et al.* Prediction of oil and gas reservoir properties using support vector machines. In *International Petroleum Technology Conference.* 2011.

[29] Nikravesh M, Adams RD, and Levey RA. Soft computing: tools for intelligent reservoir characterization (IRESC) and optimum well placement (OWP). *Journal of Petroleum Science and Engineering.* 2001;29(3–4): 239–262.

[30] Hegde C and Gray K. Use of machine learning and data analytics to increase drilling efficiency for nearby wells. *Journal of Natural Gas Science and Engineering.* 2017;40:327–335.

[31] Hampson DP, Schuelke JS, and Quirein JA. Use of multiattribute transforms to predict log properties from seismic data. *Geophysics.* 2001;66(1): 220–236.

[32] Anifowose F, Labadin J, and Abdulraheem A. Improving the prediction of petroleum reservoir characterization with a stacked generalization ensemble model of support vector machines. *Applied Soft Computing.* 2015;26: 483–496.

[33] Keith Myers REP. Time to invest or to panic? *GeoExpro: Geoscience and Technology Explained.* 2015;12(2):50–52.

[34] Chawathe A, Ouenes A, Weiss W, *et al.* Interwell property mapping using crosswell seismic attributes. In: *SPE Annual Technical Conference and Exhibition.* Society of Petroleum Engineers. 1997.

[35] Ramana M and Ramprasad T. Gas hydrate occurrence in the Krishna–Godavari offshore basin off the east coast of India. *Exploration & Production.* 2010;8(1), 22–28.

[36] Krishna Godavari Basin. *National Data Repository, Ministry of Petroleum and Natural Gas*, Government of India. 2015.

[37] McKinley S and Levine M. *Cubic spline interpolation. College of the Redwoods.* 1998;45(1):1049–1060.

[38] Chaki S, Routray A, and Mohanty WK. A novel preprocessing scheme to improve the prediction of sand fraction from seismic attributes using neural networks. *IEEE Journal of Selected Topics in Applied Earth Observations and Remote Sensing.* 2015;8(4):1808–1820.

[39] Smith SW. *The Scientist and Engineer's Guide to Digital Signal Processing.* San Diego, CA: California Technical Publishing; 1997.

[40] Fu CL, Xiong XT, and Qian Z. Fourier regularization for a backward heat equation. *Journal of Mathematical Analysis and Applications.* 2007;331 (1):472–480.

[41] Ahmad M and Sundararajan D. A fast algorithm for two dimensional median filtering. *IEEE Transactions on Circuits and Systems.* 1987;34(11):1364–1374.

[42] Valenti JCAF. Porosity prediction from seismic data using multiattribute transformations, N Sand, Auger Field, Gulf of Mexico. Gulf of Mexico, MSc Thesis, The Pennsylvania State University; 2009.

[43] Chopra S and Marfurt KJ. Seismic attributes: a historical perspective. *Geophysics.* 2005;70(5):3SO–28SO.

[44] Sánchez-Maroño N, Alonso-Betanzos A, and Tombilla-Sanromán M. Filter methods for feature selection–a comparative study. In: *International Conference on Intelligent Data Engineering and Automated Learning.* 2007; p. 178–187.

[45] Guyon I and Elisseeff A. An introduction to variable and feature selection. *Journal of Machine Learning Research.* 2003;3:1157–1182.

[46] Estévez PA, Tesmer M, Perez CA, *et al.* Normalized mutual information feature selection. *IEEE Transactions on Neural Networks.* 2009;20(2):189–201.

[47] Rocke DM and Woodruff DL. Identification of outliers in multivariate data. *Journal of the American Statistical Association.* 1996;91(435):1047–1061.

[48] De Maesschalck R, Jouan-Rimbaud D, and Massart DL. The Mahalanobis distance. *Chemometrics and Intelligent Laboratory Systems.* 2000;50(1):1–18.

[49] Drumond DA, Rolo RM, and Costa JFCL. Using Mahalanobis distance to detect and remove outliers in experimental covariograms. *Natural Resources Research.* 2019;28(1):145–152.

[50] Ekiz M and Ekiz OU. Outlier detection with Mahalanobis square distance: incorporating small sample correction factor. *Journal of Applied Statistics.* 2017;44(13):2444–2457.

[51] Mahalanobis P. On the generalized distance in statistics. *Proceedings of the National Institute of Science (India).* 1936;12:49–55.

[52]   van der Maaten L and Hinton G. Visualizing data using t-SNE. *Journal of Machine Learning Research.* 2008;9:2579–2605.

[53]   Abhishek K, Singh M, Ghosh S, *et al.* Weather forecasting model using artificial neural network. *Procedia Technology.* 2012;4:311–318.

[54]   Anifowose FA, Labadin J, and Abdulraheem A. Ensemble model of non-linear feature selection-based Extreme Learning Machine for improved natural gas reservoir characterization. *Journal of Natural Gas Science and Engineering.* 2015;26:1561–1572.

[55]   Ahmadi MA and Chen Z. Analysis of gas production data via an intelligent model: application to natural gas production. *First Break.* 2018;36(12):91–98.

[56]   Wong KW, Fung CC, Ong YS, *et al.* Reservoir characterization using support vector machines. In: *International Conference on Computational Intelligence for Modelling, Control and Automation and International Conference on Intelligent Agents, Web Technologies and Internet Commerce.* 2005, vol. 2. pp. 354–359.

[57]   Nikravesh M and Aminzadeh F. Past, present and future intelligent reservoir characterization trends. *Journal of Petroleum Science and Engineering.* 2001;31(2–4):67–79.

[58]   Pedregosa F, Varoquaux G, Gramfort A, *et al.* Scikit-learn: Machine learning in Python. *Journal of Machine Learning Research.* 2011;12(Oct):2825–2830.

[59]   Spiess AN and Neumeyer N. An evaluation of R^2 as an inadequate measure for nonlinear models in pharmacological and biochemical research: a Monte Carlo approach. *BMC Pharmacology.* 2010;10(1):6.

[60]   Zou H and Hastie T. Regularization and variable selection via the elastic net. *Journal of the Royal statistical Society: Series B (Statistical Methodology).* 2005;67(2):301–320.

[61]   Long JS and Trivedi PK. Some specification tests for the linear regression model. *Sociological Methods & Research.* 1992;21(2):161–204.

[62]   Bengio Y. Learning deep architectures AI. *Foundations and Trends® in Machine Learning.* 2009;2(1):1–127.

[63]   Mhaskar H, Liao Q, and Poggio T. When and why are deep networks better than shallow ones? *In: Thirty-First AAAI Conference on Artificial Intelligence.* 2017. pp. 2343–2349.

[64]   Liang S and Srikant R. Why deep neural networks for function approximation? In: ICLR; 2017.

[65]   Srivastava N, Hinton G, Krizhevsky A, *et al.* Dropout: a simple way to prevent neural networks from overfitting. *The Journal of Machine Learning Research.* 2014;15(1):1929–1958.

[66]   Ioffe S and Szegedy C. Batch normalization: accelerating deep network training by reducing internal covariate shift. In: *Proceedings of the 32nd International Conference on Machine Learning.* 2015, vol. 37. pp. 448–456.

[67]   He H and Garcia EA. Learning from imbalanced data. *IEEE Transactions on Knowledge & Data Engineering.* 2008;21(9):1263–1284.

Part III

Tools and technologies for Earth Observation data

Chapter 11

The application of R software in water science

*Nasrin Fathollahzadeh Attar[1] and
Mohammad Taghi Sattari[2]*

Nowadays, dealing with data in all sciences is very critical. Data science and data engineering are exciting disciplines to turn data into bright understanding. Finding statistical characteristics and communicating data by some techniques such as visualization. R is a rapidly growing, statistical, open-source software with many libraries and packages. A core team of R developers and the R Foundation for Statistical Computing support it. Recently, many scholars in different fields have used R daily and produced publication-quality graphics. Hydrologists use this software to tidy, transform, visualize, and model their hydrometric data. There are many packages available in hydrology science. Hydroinformatics is a branch of informatics that deals with water purposes. They are starting from downloading hydrological data in special packages, cleaning up hydrological and climate data, managing data such as aggregating, dealing with missing data, extracting indicators, analyzing extreme events such as floods, droughts, bushfires, dealing with data scales, spatial and temporal dataset tools, dealing with surface and groundwater, hydrographs, rainfall, snowfall water quality, reservoirs packages, watershed modelling, soil water systems, evaporation, and multiple water-related packages. This chapter aims to analyze these packages by explaining the resources and their use in water and hydrology science, finding the gaps in existing packages, and suggesting researchers develop new packages.

11.1 Introduction

Earth and environmental science, derived data technologies provide comprehensive information about the earth system. Research on environmental science has a long tradition because of continuously changing phenomena. Recent theoretical developments have revealed that data-driven methods such as machine learning (ML) and deep learning can be applied to deal with extensive environmental observations. The latter results in efficient and accurate modelling boosted by statistical, advanced cloud computing techniques. There are growing appeals for machine and

[1]Department of Statistics, University of Padova, Padua, Italy
[2]Department of Water Engineering, University of Tabriz and Ankara University, Iran

deep learning methods because of their complementary operation over classic physical-based methods.

11.1.1 What is hydrology?

The study of water is called 'hydrology', water cycle, water distribution, quantity, physical attributes, chemical and quality properties, and its relation with other beings in the earth system. The study of hydrology and its related issues is called 'hydrological sciences'. Hydrological science is an interdisciplinary field with significant challenges due to diverse data, including meteorological data, river flow, rainfall, snow data, run-off, climate change variables, soil data, soil moisture, watershed data, and others. The abundance of data in this field has led to significant achievements in various issues, thus necessitating certain requirements for the study of 'computational hydrology'.

11.1.2 What is computational hydrology?

The computational hydrology group developed tools to simulate and investigate the terrestrial hydrological cycle and applied them to many hydrologic research projects. These scientists conduct long-term monitoring and forecasting studies for droughts and streamflow, simulate the interactions between climate system components in coupled regional climate models and develop and analyze big data sets related to climate change. The computational hydrology group develops simulation and prediction tools. Ensemble forcing data are available for large-domain hydrologic models, multiscale hydrologic models, continental-domain network routing models, ensemble methods for data assimilation, and model benchmarking and evaluation methods based on the process. The work on model development targets applications such as streamflow forecasting, water security assessments, and improving model representations of hydrologic processes. Computational hydrology is a discipline that helps researchers to get data, do some preprocessing and data cleaning, analyses big data, model them in specific workflows, document them, visualize and share them in specific web-based clouds such as Hydro Share [1], hydroclient [2] and hydro server [3,4]. Hydro share is a system in which hydrologists can share data, models, and hydrologic-based resources in which this information is sharable and citable. The Consortium of Universities developed this web-based service for the Advancement of Hydrologic Sciences Incorporated (CUAHSI), originated by the United States National Science Foundation (NSF) grant [5]. The main aim of computational hydrology is to find hydrological analysis that can be applied to watersheds to properly understand the hydrological cycle. This term is different from 'Hydroinformatics'.

11.1.3 What is hydroinformatics?

Hydroinformatics primarily focuses on black-box methods to address water-related problems using famous data-driven techniques. Hydro informatics consists of two words, 'Hydro' and 'information', which carry us into the water world and give us knowledge and information using some tools and technologies to present [6]. This field consists of different collections of topics such as developing new water-related tools, developing smart water networks, complex water system management, hydrological modelling systems, smart sewage system, modelling COVID-19 reflection on the water, numerical modelling, integration of different models, big data knowledge, water data analysis, water-related data interpretation. From this point of view, the new term 'Data science' and its frameworks come to mind. Data science or data engineering is a field that utilizes different methods and statistical algorithms, and techniques to extract information insights from historical data and applies the results for future forecasting and predicting in different applications. Data science has received substantial interest in recent years and has been applied to all data types by researchers in different fields of study. In hydrology, data science plays a primary role, especially in subfields such as floods, precipitation, groundwater, water quality modelling, and prediction [7]. In addition, to apply different data science algorithms, programming languages are essential for computational hydrology and hydroinformatics. Digital water systems, innovative, sustainable water cities, and digital water networks are all applications of hydroinformatics.

11.1.4   Free, open-source software (FOSS)

Different programming languages are used in hydrology, such as Fortran, C, C++, Java, MATLAB®, Python, R, and Julia. R and Python are free open-source software (FOSS) from all the above programs. Python appeared in 1991 by Guido van Rossum in the Netherlands [8]. This site proposed the application of Python in hydrological sciences by presenting some important packages [9]. Some of the important python libraries in hydrology are related to data collection, catchment hydrological models, meteorological tools, unsaturated zones, groundwater-related, time series analysis packages, GIS and spatial related and optimization, uncertainty, and statistical packages. More information can be obtained from the 'GitHub' site [10].

11.1.5   What is GitHub?

GitHub is a platform for developing based on Git software. GitHub is a code-presenting website that is called a version control system. Many open-access software users use this site to share their projects and codes. There are millions of repositories on this site, each related to specific projects created by developers. Each repository can have its URL link to share outside of GitHub with other large communities. Each developer has a profile, and all the previous repositories and projects, people can be found on their profiles [11].

11.2   Material and methods

11.2.1   What is R? What is an integrated development environment (IDE)?

R software is a free, versatile, open-source programming language used by researchers for solving data analysis issues. It is ubiquitous in different fields of science. R is most famous for some features, including visualization (beautiful and rich set of graphs with more creative potential than Microsoft Excel), reproducibility (works with big data, building reports), advanced modelling (advanced growing statistical models), automation (model computations, cloud environments), generating reports, dashboards, and web applications. R software can be downloaded from the CRAN website [12]. One of the essential things about R, because of its open access and a significant number of users in all fields, is that there are lots of websites and information on the web available about R. But sometimes, it is hard to find them using Google search. Therefore, a search engine named 'Rseek' [13] makes it easy to search for R. Sasha Goodman and some volunteers have maintained this engine since 2007. Another similar search engine searches R in multi-sites [13].

Some integrated development environments (IDEs) software facilitates using R. IDEs have different windows simultaneously, which helps the users code. They consist of a code script window, code compiler or console, debugging options, data environment, and a single graphical user interface (GUI) for visualization of plots, help window, package installation, updating the installed packages, and documents and vignettes. The most popular IDE for R is RStudio, founded by Joseph J. Allaire in 2009, and in 2011 RStudio IDE for R was launched. The RStudio is extremely user-friendly by providing a helpful environment for all researchers. RStudio is available in an open-source edition and can be downloaded freely for use by desktops for all platforms such as Windows, Linux, or MacOS. Utilizing RStudio can help us conduct various tasks such as creating different code scripts, version control, creating presentations in Rmarkdown in different formats (HTML, Latex, PowerPoint), creating fruitful graphs, and creating web applications in shiny. More information about RStudio can be found at [14].

11.2.2 What are R packages?

R packages or R extensions include documentation (extra explanations), sample data, codes, and vignettes. All are integrated into (the Comprehensive R archive network) CRAN repository. Based on data from 28 February 2022, there are 18,980 active packages and 10,594 maintainers on CRAN [15]. The rOpenSci project [16] is a group of volunteer experts who review R packages before uploading packages to CRAN. The site repackages [17] track the number of downloads for each package. Download stats of R packages can be found at [18].

11.2.3 What are cheatsheets?

Cheatsheets, also called 'reference cards', are respective PDF versions of some famous packages that are made using packages easily by summarizing all important parts and functions in just a few pages. All the cheat sheets can be found at [19].

11.2.4 What are R communities?

Because of extensive R users worldwide and the fast pace of R, many communities support beginners (learning the materials) and professionals (solidifying their understanding). These communities strongly encourage us to join the discussions, events, and conferences for R-related topics by providing ample and rich blogs, forums, and websites. R-bloggers is a good blog updated daily and has many authors [20]—on Twitter, using the hashtag #rstats and R tip of the day [21] to get a handful and widely used R-related topics. The Revolution Analytics blog has contributed to R development topics [22]. Another online community that helps beginners to experts in question and answering platforms is StackOverflow [23]; to comment, one should be signed into the website. Another similar site to StackOverflow is a cross-validated site that helps R users with statistics, statistical models, and data mining techniques [24]. RStudio also has a huge community of R-users and developers [25]. Their community also has some mailing lists for getting help using R [26]. These mailing lists have been active since the 1990s. Weekly also contains learning R materials, blog posts, podcasts, and the newest updates [27]. R meet-up groups are also overgrowing [28], with 38 countries, 92 groups, and approximately 72,101 members until the 28th of February 2022. Another global community that promotes gender diversity is the R users community, which has many active branches from all over the world, with 140 groups and 79,251 members to the date of this study (2022.02.28) [29]. R girls school is also a network that promotes the R for secondary schools (11–16 years old) and is currently developing lesson plans for use in the classrooms [30]. Lastly, R communities, developers, users, distributors, maintainers, and R-related conferences are supported by R Consortium [31].

11.2.5   What is RPubs?

RPubs [32] is an open-source place for publishing presentations and scientific resources created in R Markdown. R Markdown documents or presentations are simply made within RStudio, and by clicking the 'Publish button'; files can be published for the readers and their feedback.

11.2.6   What are popular conferences in R?

Some R-related conferences are organized regularly by members of R communities.

NICAR 2022 (March 3–6), the Investigative Reporters & Editors Conference, should be well attended by data journalists using R.

Appsilon Shiny Conference (April 27–29) will bring together members from the global community of Shiny developers to learn, network, and collaborate.

R/Finance (June 3–4) is the primary meeting for academics and practitioners interested in using R in quantitative finance.

R CoR Conference (June 8–10)

useR! 2022 (June 20–23) remains virtual to make the conference accessible and inclusive in as many ways as possible.

rstudio::conf 2022 (July 25–28)

BioC 2022 (July 27–29) showcases the use of an open-source in bioinformatics.

JSM 2022 (August 6–11), one of the largest statistical events in the world, will likely have several R-related talks in its program.

R Medicine (August 23–26) promotes using R-based tools to improve clinical research and practice.

EARL Conference (September 6–8) is a cross-sector conference focusing on the commercial use of the R programming language.

R Government (December, dates TBA)

Past events

Evidence synthesis and meta-analysis in R—The talks from this workshop series aim to develop and promote open software tools for evidence synthesis.

11.2.7   What is joss (open source software)?

The Joss stands for Journal of Open Source Software, a peer-reviewed, developer-friendly, open-access scientific journal covering open-source software from any research field in any programming language. The journal uses GitHub as a publishing platform [33]. Based on the site information at the time of writing this study (data issued on February 28, 2022), there are 1,748 published papers and 1,555 active papers. The papers can be searched by the title, tag, author, or language in the search button.

11.2.8   What is R studio cloud?

RStudio Cloud [34] has recently been developed to simplify data science for professionals, trainers, teachers, and students, especially in hydrological sciences. It can be challenging to strike a balance between teaching students how to program while not ignoring the hydrologic principles at the center of a class. Rstudio cloud provides a controlled environment for students to learn programming principles, where they can make minor adjustments to pre-written scripts and see how they work. Students can work on their assignments in the cloud from any computer with an Internet connection. It also features interactive tutorials describing the basics of data science, cheatsheets for working with popular R packages, a guide to using RStudio Cloud, and links to DataCamp® courses.

11.2.9   What is R application in hydrology?

R programming language holds interest among the hydrologist community, and this causes them to develop useful and efficient hydrology packages, workflows, hydrology R-related events, conferences, and holding training courses such as the application of R in hydroinformatics. Figure 11.1 represents the word cloud of repeated terms of hydrology in R-reviewed papers in this study. R is used for high-performance computing of hydrological observations and big data. The applications of R in hydrology can be folded into (1) getting hydrological data and visualization, (2) developing apps and packages for hydrological data, and (3) reporting and summarizing hydrological models.

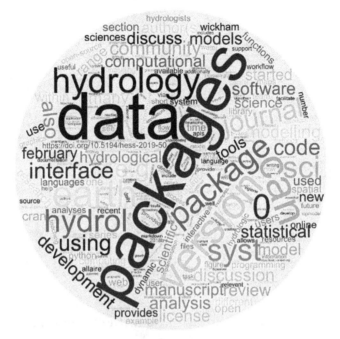

Figure 11.1   The word cloud of using R in hydrology

11.2.10   What are hydrological packages?

R packages are also growingly used in hydrological studies such as surface water, groundwater quantity and quality, meteorological data, data tiding packages (gap-filling, data cleaning, missing data, organization), hydrograph analysis (function to work with flow data, flow trends, flow statistics and indices), spatial data and GIS application packages, statistical modelling and other packages. The whole list can be obtained here [35]. All packages which are utilized in R are citable. The code below shows how to refer to a package in a publication by giving the title, author's name, journal name, volume, number, doi, and URL of the package, which can be added in reference manager software such as Mendeley or endnote.

Code: writeLines(toBibtex(citation('package name')))

11.2.11   Workflow of R in hydrology

To have a reproducible computational hydrological model in R; these stages should be presented in every model: (1) retrieving data sets; (2) preprocessing scripts; (3) defining input parameters; (4) splitting data sets into calibration and validation stages; (5) model building; (6) reporting the results in more innovative ways. Figure 11.2 symbolizes different stages of the R in hydrology workflow.

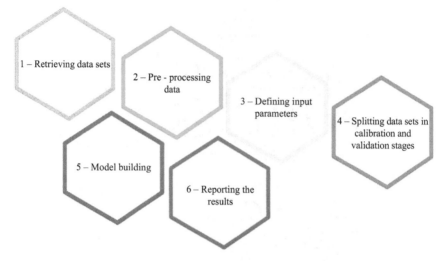

Figure 11.2   The flowchart of workflow in R

11.2.12   Data for hydrology? How to retrieve datasets?

The rise of large-scale data archives hydro-climatological big data along with spatial and satellite data available in the R environment for retrieval makes R incredibly unique programming software for researchers in hydrological studies. All hydrology-related data such as surface water (river flow), meteorological and climate data, satellite data, snow data, and precipitation can be downloaded from data retrieval packages within R. Increasingly, researchers are employing data science to study hydrology. Insights into process behavior can be gained from large and complex data sets. Table 11.1 shows different data retrieval packages for hydrology purposes.

11.2.13   Preprocessing retrieved hydrological data (data tidying)

Data are generated at high speed every second; data quality is almost inevitable in creating accurate models. Clean data is the first step to having precise models. Data tidying arguably is the most important part of the hydrology workflow [65]. If data has fluctuations, anomalies, outliers, gaps, and duplications, it will make wrong models and decisions. Thus R has some useful libraries and packages that clean data before modelling, which are demonstrated in Table 11.2.

11.2.14   Different hydrology model types?

Hydrological models can be subdivided into two main categories of physical-based models and abstract (deterministic or mathematical) models, as confirmed in Figure 11.3. Physically based models consider the physical characteristics of watersheds or processes of the water cycle. On the other hand, abstract models are

Table 11.1   List of packages for data retrieval

Package name	Usage	Details
AWAPer	Catchments (example catchment boundary polygons)	Catchment Area Weighted Climate Data Anywhere in Australia [36]
dataRetrieval	Download water data	Water quality and hydrology data from EPA and USGS [37]
echor	Download discharge records	Provides functions to locate facilities with discharge permits and download discharge records [38]
FedData	Downloading geospatial data	Available from several federated data sources [39]
hydroscoper	R interface to the Greek National Data Bank for Hydrological and Meteorological Information	Functions to transliterate, translate, and download them into tidy data frames (tibbles) [40]
metScanR	A tool for locating, mapping, and gathering environmental data	Metadata from over 157,000 environmental monitoring stations among 219 countries/territories [41]
nhdR	National Hydrography Dataset. Technical Report. United States-Geological Survey	Functions for querying, downloading, and networking [42]
rnrfa	Generate a map and extract time series and general information	Functions to retrieve data from the UK National River Flow Archive [43]
tidyhydat	Historical and real-time national 'hydrometric' data	Provides functions to access historical and real-time national 'hydrometric' data from Water Survey of Canada data sources [44]
washdata	Urban water and sanitation survey dataset from the survey conducted in Dhaka, Bangladesh	Part of a series of surveys to be conducted in various cities, including Accra, Ghana; Nakuru, Kenya; Antananarivo, Madagascar; Maputo, Mozambique; and Lusaka, Zambia [45]
waterData	US Geological Survey (USGS) daily hydrologic data	Plots the data, addresses some common data problems, and calculates and plots anomalies [46]
climate	Atmospheric vertical profiling data, and Polish Institute of Meteorology and Water Management	Downloading of meteorological and hydrological data from publicly available repositories [47]
clifro	New Zealand National Climate Database	Website of New Zealand National Climate Database of around 6,500 climate stations [48]
getMet	Meteorological data for hydrologic models	The ability to source, format, and edit meteorological data for hydrologic models [49]
GSODR	Global Surface Summary of the Day (GSOD) weather data from USA	Automated downloading, parsing, cleaning, unit conversion and formatting [50]

(Continues)

Table 11.1   (Continued)

Package name	Usage	Details
MODISTools	MODIS Land Products Subset	Allows for easy downloads of 'MODIS' time series [51]
nasapower	Worldwide Energy Resource (POWER) project data in API	Daily meteorology, interannual and 30-year climatology [52]
metR	Handling meteorological data	Commonly used analysis methods in the atmospheric sciences [53]
prism	Access and visualize data from the Oregon State PRISM project	Data are presented as gridded rasters at four different temporal scales: daily, monthly, annual, and 30 years normal [54]
rdwd	Climate data from the German	Download observational time series from meteorological stations [55]
RNCEP	NCEP/NCAR Reanalysis and NCEP/DOE Reanalysis II datasets	Functions to retrieve, organize, and visualize weather data [56]
rnoaa	NOAA data sources in API	Data, data sets, types, locations, locations, and stations. Includes interfaces for NOAA sea ice data, severe weather inventory, historical Observing Metadata Repository ('HOMR'), storm data via 'IBTrACS', and tornado data via the NOAA storm prediction center [57]
rpdo	Monthly Pacific Decadal Oscillation (PDO) index values	Downloading Southern Oscillation Index, Oceanic Nino Index, and North Pacific Gyre Oscillation data [58]
rwunderground	Temperature, humidity, wind chill, wind speed, dew point, and heat index	Getting historical weather information and forecasts from wunderground.com [59]
smapr	NASA Soil Moisture Active-Passive (SMAP) Data	Acquire and extract NASA Soil Moisture Active Passive (SMAP) data [60]
stationaRy	A global network of weather stations provides hourly weather data	A 'tibble' is automatically generated for the exact range of years requested from a data repository [61]
worldmet	Download data from world meteorological sites	Over 30,000 surface meteorological sites around the world are imported [62]
cdlTools	National Agricultural Statistics Service data	Crop shape data for a specified state [63]
FAOSTAT	Agricultural statistics provided by the FAOSTAT	Food and Agricultural Organization of the United Nations database [64]

Table 11.2   List of packages for data tidying

Package name	Usage	Details
driftR	Cleaning and correcting water quality data	Open-source data sets (ex: USGS and NDBC), all of which are susceptible to errors/inaccuracies due to drift [66]
climdex.pcic	PCIC implementation of Climdex Routines	Computation of extreme climate indices [67]
climatol	Functions to quality control, homogenization, and missing data infilling of climatological series	Climate diagrams of Walter and Lieth drawn [68]
plyr	Visualization of data	Create histogram [69]
tidyr	Tidy your data	Cleaning your data by identifying the variables in your dataset and using the tools provided [70]
janitor	Cleaning dirty data	Able to find duplicates by multiple columns [71]
splitstackshape	Work with comma-separated values in a data frame column	Useful for survey or text analysis preparation [72]
hyfo	Data processing and visualization in hydrology	Data of precipitation extracted, data downscaled and resampled, and get filled with this package [73]

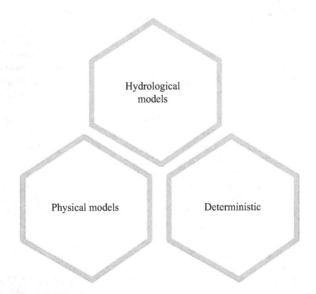

Figure 11.3   Different types of hydrological modeling

used to demonstrate all hydrological parameters in algebra mode. The deterministic equations, input, and output result from mathematical models. Using these models, the researchers can predict or forecast hydrological issues, extreme events, or future climate fluctuations. The utilized variables for these models are functions of space and time.

Hydrological-R users use and developed some packages for both physical-based and deterministic models. Table 11.3 shows some of the essential packages with their details obtained from the CRAN repository.

Table 11.3　List of packages for physically based and deterministic models

Physically based models		Deterministic (statistical) models	
Package name	Details	Package name	Details
airGR	Conceptual rainfall-runoff models and snow accumulation and melt model [74]	CoSMoS	Generates univariate/multi-variate non-Gaussian time series [75]
bigleaf	Calculation of physical and physiological ecosystem properties [76]	hydroApps	Regional analysis of hydrological applications [77]
boussinesq	Boussinesq equation (ground-water model-ling) [78]	hydroGOF	Goodness-of-fit measures between observed and simulated values [79]
Ecohydmod	Soil water balance simulation [80]	HydroMe	Parameters in infiltration and water retention models [81]
EcoHydRology	SWAT calibration functions [82]	LPM	Long Memory Models to hydrological data sets [83]
geotopbricks	Hydrological distributed model GEOtop [84]	nsRFA	Regional frequency analysis methods in hydrology [85]
hydroPSO	Particle swarm optimisation (PSO) algorithm for the calibration of environmental models [86]	RMAWGEN	Stochastic generation of daily time series of temperature and precipitation [87]
HBV.IANI-GLA	The HBV hydrological model [88]	SCI	Functions for generating Standardized Climate Indices [89]
kwb.hantush	Calculation of ground-water mounding beneath an infiltration basin [90]	Soil water	Soil water retention or conductivity curve [91]
RavenR	Raven hydrological modeling framework [92]	Synthesis	Generate synthetic time series [93]

(Continues)

Table 11.3 (*Continued*)

Physically based models		Deterministic (statistical) models	
Package name	Details	Package name	Details
reservoir	Analysis, design, and operation of water supply storages [94]	SPEI	Standardized Precipitation—Evapotranspiration Index (SPEI) [95]
RHMS	Construction, simulation, visualization, and calibration of hydrologic systems [96]	WASP	A wavelet-based variance transformation method [97]
RSAlgaeR	Empirical remote sensing models of water quality variables [98]	Evapotranspiration	Functions to calculate potential evapotranspiration (PET) and actual evapotranspiration (AET) [99]
streamDepletr	Calculate the impacts of groundwater pumping [100]	MBC	Multivariate bias correction of climate model output [101]
swmmr	Storm water management model (SWMM) [102]	meteoland	Functions to estimate weather variables [103]
telemac	Modeling of free surface flow [104]	musica	Multiscale Climate Model Assessment [105]
topmodel	Hydrological functions TOPMODEL [106]	openair	Tools to analyze, interpret, and understand air pollution data [107]
TUWmodel	Lumped hydrological model for education purposes [108]	qmap	Climate model simulations using quantile mapping [109]
WRSS	Water resources system simulator [110]	MODIStsp	MODIS satellite data can be downloaded along with preprocessing Land Products Data [111]

11.2.15   *Hydrologic time series analysis tools in R?*

The hydrologic time series analysis plays an essential role in water resource planning and management. Statistical analyses of every hydrologic time series are required to determine fundamental characteristics of normality, homogeneity, stationarity, trends and shifts, periodicity, persistence, and a stochastic component. However, a similar practice is absent or less in hydrology and hydrogeology. Thus, hydrologic time series analysis has received relatively little attention, even in the age of information technology. It is notably lacking in water resources engineering to deal with both theory and application of time series analysis techniques. Thus, time series analysis is not adopted easily by many hydrologists and hydrogeologists. But, relevant packages in R software make it easy to work with time-series data and model them for future prediction or forecasting. Table 11.4 demonstrates useful tools and packages obtained from the CRAN task view [112].

Table 11.4   List of packages for time series analysis

Package name	Usage	Details
fable	Tools for fitting univariate time series models	ETS, ARIMA, TSLM, and other models [113]
forecast	Provides time series forecasting tools	Functions for computing and analyzing forecasts [114]
prophet	Analyzes time series using an additive model that fits yearly and weekly seasonality to nonlinear trends	Works best with daily data [115]
tseries	Time series analysis and computational	Fits basic GARCH models [116]
tsDyn	Nonlinear time series models with regime switching	Implements nonlinear autoregressive (AR) time series models [117]
MTS	All-purpose toolkit for analyzing multivariate time series	VAR, VARMA, seasonal VARMA, VAR models with exogenous variables, and more [118]
TSrepr	Methods for representing time series	Using dimension reduction and feature extraction [119]
wavelets	Wavelet methods	Computing wavelet filters, wavelet transforms, and multiresolution analyses [120]

11.2.16   Hydrological ML application tools in R?

ML, which is growing rapidly, offers methodological opportunities compatible with hydrological research needs and challenges. The era of big data in hydrology has undoubtedly arrived with the expansion of measurement networks, more frequent automatic measurements of hydrological variables, and the increased use of remote sensing products. Models based on processes are typically developed for specific spatio-temporal scales that are difficult to adjust to new datasets. Numerous applications have demonstrated the superiority of automatic methods that recognize patterns and generalize them. A training dataset contains calibration and validation data in most ML techniques. A simple example of a bias-variance tradeoff will be discussed since these data are usually spatially and temporally correlated. From artificial neural networks, and Support Vector Machines (SVMs), to gradient boosting machines, we present the ML algorithms roughly following chronological order. The importance of these and other ML techniques will grow in hydrology as data streams increase. Table 11.5 [121] shows some of the important ML packages available in R. List of ML packages in R for remote sensing applications are displayed in Table 11.6.

11.2.17   Remote sensing tools in R

This tool deals with both theory and application of time series analysis techniques. Thus, time series analysis is not adopted easily by many hydrologists and

Table 11.5   List of packages for ML algorithms

Package name	Usage	Details
nnet	Neural networks package	Single-hidden-layer neural networks are implemented [122]
deepnet	Deep learning package	Feed-forward neural network, restricted Boltzmann machine, deep belief network, stacked autoencoders [123]
rpart	Tree-structure CART model	Classification and survival analysis [124]
RWeka	Tree-structure models	J4.8-variant of C4.5 and M5 implementation [125]
randomForest	Regression and classification	Implementation of the random forest algorithm [126]
xgboost	Boosting	Tree-based boosting using efficient trees as base learners for several and also user-defined objective functions [127]
rgenoud	Optimization using Genetic Algorithms	Offers optimization routines based on genetic algorithms [128]
frbs	Fuzzy rule-based system	Regression and classification using Fuzzy technique [129]

Table 11.6   List of packages for remote sensing applications

Package name	Usage	Details
RStoolbox	Toolbox for remote sensing image processing and analysis	Calculating spectral indices, principal component transformation, unsupervised and supervised classification, or fractional cover analyses [132]
landsat	Radiometric and topographic correction of satellite imagery	Includes relative normalization, image-based radiometric correction, and topographic correction options [133]
hsdar	Manage, analyze and simulate hyperspectral data	Transformation of reflectance spectra, calculation of vegetation indices and red edge parameters, spectral resampling for hyperspectral remote sensing, simulation of reflectance and transmittance using the leaf reflectance model PROSPECT and the canopy reflectance model PROSAIL [134]
rasterVis	Visualization methods for raster data	Methods for enhanced visualization and interaction with raster data. It implements visualization methods for both quantitative and categorical data for univariate and multivariate rasters. It also provides methods to display spatiotemporal rasters and vector fields [135]

hydrogeologists. But, relevant packages in R software make it easy to work with time-series data and model them for future prediction or forecasting.

Remote sensing is most frequently called collecting information about the Earth and other planets. This term is used in various fields, including geography, land surveying, and most of Earth's scientific fields such as hydrology, ecology, meteorology, oceanography, glaciology, and geology [130]. Monitoring data with wide geographic coverage derived from modern remote sensing methods are now reasonably priced and readily accessible [131]. In recent years, the technical advancements of drone images, aerial orthophotos, and satellite remote sensing have been substantial and rapid.

For the evaluation of water resources, very accurate water body mapping is required, and quick water body mapping is needed for flood monitoring. Large-scale water body mapping is particularly good for synthetic aperture radar (SAR), which collects data in all lighting and weather scenarios. Sentinel-1's excellent temporal–spatial resolution allows for precise water body monitoring. The SVM classifier separates water from non-water to recognize surface water from time series Sentinel-1 data. Then, using the Random Forest Regressor (RFR), the value of the surface water may be forecasted for the gap period or the time when there is no data. The fusion of remote sensing time series data and ML methods in an R environment can increase the reliability of surface water detection for flood mapping.

11.3   Conclusion and future prospects

In recent years, R programming has gained much attention in all fields of study. It also acquired an essential role in remote sensing, hydrology, and hydrological research, focusing on the operational practice of hydrology by facilitating a wide range of hydrological analyses. In this chapter, we have restricted hydrology-related packages in R software and have introduced them. In this study, the authors' utilized resources such as CRAN task view in hydrology, GitHub site, and R-Forge website (last accessed on February 28, 2022) to gain information about hydrology-related packages. R has many tools for visualization, modelling (deterministic or physical based), spatial modelling, statistical computation, data retrieval, and pre and post-processing in hydrological sciences. The increasing use of computational hydrology, hydroinformatics in water environment-related fields can play a crucial role in growing the R users' community, and future advances and development of hydrological-related packages. The code submitting and sharing with other r-hydro communities in paper submission will be another prospect for the future. The teaching of this open-source programming will be continued in schools, universities, and communities more often in the near future.

References

[1]   Hydroshare. Available from: https://www.hydroshare.org/
[2]   Hydroclient. Available from: https://data.cuahsi.org/

[3]   Essawy BT, Goodall JL, Zell W, *et al.* Integrating scientific cyberinfras-
      tructures to improve reproducibility in computational hydrology: example
      for HydroShare and GeoTrust. *Environ Model Softw.* 2018;105:217–229.
      Available from: https://doi.org/10.1016/j.envsoft.2018.03.025

[4]   hydro server. Available from: https://hydroserver.cuahsi.org/

[5]   Tarboton DG and Idaszak R. HydroShare: advancing hydrology through
      collaborative data and model sharing. *iRODS User Gr Meet.* 2015. Available
      from: https://irods.org/documentation/articles/irods-user-group-meeting-2015/,
      https://irods.org/wp-content/uploads/2015/06/Tarboton-HydroShare.pdf.

[6]   Abbott MB. Introducing hydroinformatics. *J Hydroinformatics.* 1999;1
      (1):3–19.

[7]   Mosaffa H, Sadeghi M, Mallakpour I, Naghdyzadegan Jahromi M, and
      Pourghasemi HR. Application of machine learning algorithms in hydrology.
      In: Pourghasemi HR, (eds.) *Computers in Earth and Environmental
      Sciences.* Elsevier. 2022. pp. 585–591 (Chapter 43). Available from: https://
      www.sciencedirect.com/science/article/pii/B9780323898614000270

[8]   Rossum G. Python programming language. In: *USENIX Annual Technical
      Conference.* 2007. p. 36.

[9]   AGU. Available from: https://agu-h3s.org/2021/03/29/resources-for-pro-
      gramming-in-hydrology/

[10] Python-Hydrology-Tools. Available from: https://github.com/raoulcollen-
      teur/Python-Hydrology-Tools

[11] Dabbish L, Stuart C, Tsay J, and Herbsleb J. Social coding in GitHub:
      transparency and collaboration in an open software repository. In: *CSCW
      '12: Proceedings of the ACM 2012 Conference on Computer Supported
      Cooperative Work*; 2012; pp. 1277–1286.

[12] CRAN. Available from: https://cran.r-project.org/

[13] Search Engine. Available from: http://www.dangoldstein.com/search_r.html

[14] Rstudio. Available from: https://www.rstudio.com/

[15] RPackage site. Available from: https://www.r-pkg.org/

[16] ropensci. Available from: https://ropensci.org/

[17] r packages. Available from: http://www.datasciencemeta.com/rpackages

[18] Package downloads. Available from: https://cran.r-project.org/web/packa-
      ges/dlstats/vignettes/dlstats.html

[19] Cheat   sheet.   Available   from: https://www.rstudio.com/resources/
      cheatsheets/

[20] r-bloggers. Available from: http://www.r-bloggers.com/

[21] R tip of day. Available from: https://twitter.com/rlangtip

[22] R blog. Available from: https://blog.revolutionanalytics.com/

[23] Stack over flow. Available from: https://stackoverflow.com/

[24] stackexchange. Available from: https://stats.stackexchange.com/questions/
      tagged/r

[25] R studio community. Available from: https://community.rstudio.com/

[26] stat.ethz. Available from: https://stat.ethz.ch/mailman/listinfo/r-help

[27] R weekly. Available from: https://rweekly.org/

[28]   R meet Ups. Available from: https://www.meetup.com/en-AU/pro/r-user-groups

[29]   R ladies Global. Available from: https://rladies.org/

[30]   R girls. Available from: https://greenoak.bham.sch.uk/r-girls-school-network/

[31]   R Consortium. Available from: https://www.r-consortium.org/

[32]   R pubs. Available from: https://rpubs.com/

[33]   Journal of Open Source Software. Available from: https://github.com/open-journals/joss

[34]   studio cloud. Available from: https://rstudio.cloud/

[35]   Hydrology Cran. Available from: https://cran.r-project.org/web/views/Hydrology.html

[36]   T. J. Peterson, C. Wasko, M. Saft, and M. Peel. AWAPer: an R package for area weighted catchment daily meteorological data anywhere within Australia. *Hydrol Process.* 2020;34:1301–1306.

[37]   Johnson LA, De Cicco LA, Lorenz D, *et al.* dataRetrieval: R Packages for Discovering and Retrieving Water Data Available from U.S. Federal Hydrologic Web Services. *US Geol Survey*; 2021.

[38]   Schramm M. *{echor}: Access EPA "ECHO" Data*; 2020.

[39]   Kyle Bocinsky R. FedData: Functions to Automate Downloading Geospatial Data Available from Several Federated Data Sources. 2019. Available from: https://cran.r-project.org/package=FedData

[40]   Vantas K. hydroscoper: R interface to the Greek National Data Bank for Hydrological and Meteorological Information. *J Open Source Softw.* 2018;3 (23):625.

[41]   Roberti J, Flagg C, Stanish L, Lee R, Weintraub S, and Smith D metScanR: Find, Map, and Gather Environmental Data and Metadata. 2017. Available from: http://cran.nexr.com/web/packages/metScanR/index.html

[42]   USGS. National Hydrography Dataset. Technical Report. United States Geological Survey. 2022. Available from: https://www.usgs.gov/national-hydrography

[43]   Vitolo C, Fry M, Buytaert W, Spencer M, and Gauster T. rnrfa: an R package to retrieve, filter and visualize data from the UK National River Flow Archive. *R. J.*, 2016:8(2):102–116.

[44]   Albers S. tidyhydat: Extract and Tidy Canadian Hydrometric Data. J Open Source Softw. 2017. Available from: http://dx.doi.org/10.21105/joss.00511

[45]   Guevarra E. washdata: Urban Water and Sanitation Survey Dataset. 2020. Available from: https://cran.r-project.org/package=washdata

[46]   Ryberg KR and Vecchia AV. waterData: Retrieval, Analysis, and Anomaly Calculation of Daily Hydrologic Time Series Data. 2017. Available from: https://cran.r-project.org/package=waterData

[47]   Czernecki B, Głogowski A, and Nowosad BC. Climate: an R package to access free in-situ meteorological and hydrological datasets for environmental assessment. *Sustainability.* 2020;12:394. Available from: https://github.com/bczernecki/climate/

[48] Shears BS and Nick TS. New Zealand's Climate Data in R – An Introduction to clifro. 2015. Available from: http://stattech.wordpress.fos.auckland.ac.nz/2015-02-new-zealands-climate-data-in-r-an-introduction-to-clifro/

[49] Sommerlot A, Fuka D, and Easton Z. getMet: Get Meteorological Data for Hydrologic Models. 2016. Available from: https://cran.r-project.org/package=getMet

[50] Sparks AH, Hengl T, and Nelson A. GSODR: global summary daily weather data in R. *J Open Source Softw.* 2017;2(10):177.

[51] Tuck SL, Phillips HR, Hintzen RE, Scharlemann JP, Purvis A, and Hudson LN. MODISTools – Downloading and Processing MODIS Remotely Sensed Data in R. *Ecol Evol.* 2014;4(24):4658–4668.

[52] Sparks AH. nasapower: a NASA POWER global meteorology, surface solar energy and climatology data client for R. *J Open Source Softw.* 2018;3(30):1035.

[53] Campitelli E. metR: Tools for Easier Analysis of Meteorological Fields. 2021.

[54] Hart EM, Bell K, and Butler A. prism: Download Data from the Oregon Prism Project. 2015. R package version 0.0, 6(10.5281).

[55] Boessenkool B. rdwd: Select and Download Climate Data from "DWD" (German Weather Service). 2021. Available from: https://cran.r-project.org/package=rdwd

[56] Kemp MU, Loon EEV, Baranes JS, and Bouten W. *RNCEP: Obtain, Organize, and Visualize NCEP Weather Data*, 2020. Available from https://cran.r-project.org/web/packages/RNCEP/index.html.

[57] Chamberlain S. rnoaa: "NOAA" Weather Data from R. 2021. Available from: https://cran.r-project.org/package=rnoaa

[58] Thorley J, Mantua N, and Hare SR. rpdo: Pacific Decadal Oscillation Index Data. 2020. Available from: https://cran.r-project.org/package=rpdo

[59] Shum A. rwunderground: R Interface to Weather Underground API. 2018. Available from: https://cran.r-project.org/package=rwunderground

[60] Joseph M, Oakley M, and Schira Z. smapr: Acquisition and Processing of NASA Soil Moisture Active-Passive (SMAP) Data. 2019. Available from: https://cran.r-project.org/package=smapr

[61] Iannone R. stationaRy: Detailed Meteorological Data from Stations All Over the World. 2020. Available from: https://cran.r-project.org/package=stationaRy

[62] Carslaw D. worldmet: Import Surface Meteorological Data from NOAA Integrated Surface Database (ISD). 2021. Available from: https://cran.r-project.org/package=worldmet

[63] Lisic LC and Chen L. Tools to Download and Work with USDA Cropscape Data. 2018. Available from: https://cran.r-project.org/package=cdlTools

[64] Michael CJK, Markus G, and Filippo G. FAOSTAT: Download Data from the FAOSTAT Database. 2022. Available from: https://cran.r-project.org/package=FAOSTAT

[65] Van Den Broeck J, Cunningham SA, Eeckels R, and Herbst K. Data cleaning: detecting, diagnosing, and editing data abnormalities. *PLoS Medicine*, 2005;2(10):966–971.

[66] Shaughnessy A, Prener C, and Hasenmueller E. shaughnessyar/driftR: driftR 1.1.0. 2018 Jun 13 [cited 2022 Feb 23]. Available from: https://doi.org/10. 5281/zenodo.1288819#.YhXQ1kZBc-I.mendeley

[67] Consortium DB for the PCI. climdex.pcic: PCIC Implementation of Climdex Routines. 2020. Available from: https://cran.r-project.org/package=climdex. pcic

[68] Guijarro JA. climatol: Climate Tools (Series Homogenization and Derived Products). 2019. Available from: https://cran.r-project.org/package=climatol

[69] Wickham H. The split-apply-combine strategy for data analysis. *J Stat Softw*. 2011;40(1):1–20. Available from: http://www.jstatsoft.org/v40/i01/

[70] Vaughan D, Girlich M, Ushey K, and Posit PBC. tidyr: Tidy Messy Data. 2022. Available from: https://cran.r-project.org/package=tidyr

[71] Firke S. janitor: Simple Tools for Examining and Cleaning Dirty Data. 2021. Available from: https://cran.r-project.org/package=janitor

[72] Mahto A. splitstackshape: Stack and Reshape Datasets After Splitting Concatenated Values. 2019. Available from: https://cran.r-project.org/pack-age=splitstackshape

[73] Xu Y. hyfo: Hydrology and Climate Forecasting. 2020. Available from: https://cran.r-project.org/package=hyfo

[74] Coron L, Delaigue O, Thirel G, *et al.* {airGR}: Suite of {GR} Hydrological Models for Precipitation-Runoff Modelling. R News. 2021. Available from: https://cran.r-project.org/package=airGR

[75] Papalexiou SM, Serinaldi F, Strnad F, Markonis Y, and Shook K. CoSMoS: Complete Stochastic Modelling Solution. 2021. Available from: https://cran. r-project.org/package=CoSMoS

[76] Knauer J, El-Madany TS, Zaehle S, and Migliavacca M. Bigleaf – an R package for the calculation of physical and physiological ecosystem prop-erties from eddy covariance data. *PLoS One*. 2018;13(8):e0201114. Available from: https://doi.org/10.1371/journal.pone.0201114

[77] Ganora D. hydroApps: Tools and Models for Hydrological Applications. 2014. Available from: https://cran.r-project.org/package=hydroApps

[78] Cordano EM and Rigon R. A mass-conservative method for the integration of the two-dimensional groundwater (Boussinesq) equation. *Water Resour Res*. 2013:49(2):1058–1078.

[79] Zambrano-Bigiarini M. hydroGOF: Goodness-of-Fit Functions for Comparison of Simulated and Observed Hydrological Time Series. 2020. Available from: https://github.com/hzambran/hydroGOF

[80] Souza R. Ecohydmod: Ecohydrological Modelling. 2017. Available from: https://cran.r-project.org/package=Ecohydmod

[81] Omuto CT, Maechler M, and Too V. HydroMe: Estimating Water Retention and Infiltration Model Parameters using Experimental Data. 2021. Available from: https://cran.r-project.org/package=HydroMe

[82] Fuka FD, Walter MT, Archibald JA, Steenhuis TS, and Easton ZM EcoHydRology: A Community Modeling Foundation for Eco-Hydrology. 2018. Available from: https://cran.r-project.org/package=EcoHydRology

[83] Tallerini C and Grimaldi S. LPM: Linear Parametric Models Applied to Hydrological Series. 2020. Available from: https://cran.r-project.org/package=LPM

[84] Cordano E, Andreis D, and Zottele F. geotopbricks: An R Plug-in for the Distributed Hydrological Model GEOtop. 2020. Available from: https://cran.r-project.org/package=geotopbricks

[85] Viglione A. nsRFA: Non-Supervised Regional Frequency Analysis. 2020. Available from: https://cran.r-project.org/package=nsRFA

[86] Zambrano-Bigiarini M and Rojas R. A model-independent Particle Swarm Optimisation software for model calibration. *Environ Model Softw.* 2013:43:5–25.

[87] Cordano E and Eccel E. RMAWGEN: Multi-Site Auto-Regressive Weather GENerator. 2017. Available from: https://cran.r-project.org/package=RMAWGEN

[88] Toum E. {HBV.IANIGLA}: Modular Hydrological Model. 2021. Available from: https://cran.r-project.org/package=HBV.IANIGLA

[89] Stagge JH, Tallaksen LM, Gudmundsson L, Van Loon AF, and Stahl K. Candidate distributions for climatological drought indices (SPI and SPEI). *Int J Climatol.* 2015:35(13):4027–4040.

[90] Rustler M. kwb.hantush: Calculation of Groundwater Mounding Beneath an Infiltration Basin. 2019. Available from: https://cran.r-project.org/package=kwb.hantush

[91] Cordano E, Andreis D, and Zottele F. soilwater: Implementation of Parametric Formulas for Soil Water Retention or Conductivity Curve. 2017. Available from: https://cran.r-project.org/package=soilwater

[92] Chlumsky R, Craig J, Scantlebury L, *et al.* RavenR: Raven Hydrological Modelling Framework R Support and Analysis. 2022. Available from: https://cran.r-project.org/package=RavenR

[93] Jiang Z. synthesis: Generate Synthetic Data from Statistical Models. 2021. Available from: https://cran.r-project.org/package=synthesis

[94] Turner SW and Galelli S. Water supply sensitivity to climate change: an R package for implementing reservoir storage analysis in global and regional impact studies. *Environ Model Softw.* 2016:76:13–19.

[95] Beguería S and Vicente-Serrano SM. SPEI: Calculation of the Standardised Precipitation-Evapotranspiration Index. 2017. Available from: https://cran.r-project.org/package=SPEI

[96] Araghinejad RAS. RHMS: Hydrologic Modelling System for R Users. 2021. Available from: https://cran.r-project.org/package=RHMS

[97] Jiang Z, Rashid MdM, Sharma A, and Johnson F. WASP: Wavelet System Prediction. 2022. Available from: https://cran.r-project.org/package=WASP

[98] Hansen C. RSAlgaeR: Builds Empirical Remote Sensing Models of Water Quality Variables and Analyzes Long-Term Trends. 2018. Available from: https://cran.r-project.org/package=RSAlgaeR

[99] Guo D, Westra S, and Peterson T. Evapotranspiration: Modelling Actual, Potential and Reference Crop Evapotranspiration. 2022. Available from: https://cran.r-project.org/package=Evapotranspiration

[100] Zipper SC. streamDepletr: Estimate Streamflow Depletion Due to Groundwater Pumping. 2020. Available from: https://cran.r-project.org/package=streamDepletr

[101] Cannon AJ, Sobie SR, and Murdock TQ. Bias correction of GCM precipitation by quantile mapping: how well do methods preserve changes in quantiles and extremes?. *J Clim.* 2015:28(17):6938–6959.

[102] Leutnant D, Döring A, and Uhl M. Swmmr – an R package to interface SWMM. *Urban Water J.* 2019:16(1):68–76.

[103] De Cáceres M, Nicolas Martin-St. Paul NM, Turco M, Cabon A, and Granda V. Estimating daily meteorological data and downscaling climate models over landscapes. *Environ Model Softw.* 2018;108:186–196.

[104] Pilz T. telemac: R Interface to the TELEMAC Model Suite. 2022. Available from: https://cran.r-project.org/package=telemac

[105] Hanel M. musica: Multiscale Climate Model Assessment. 2016. Available from: https://cran.r-project.org/package=musica

[106] Buytaert W. topmodel: Implementation of the Hydrological Model TOPMODEL in R. 2018. Available from: https://cran.r-project.org/package=topmodel

[107] Carslaw DC and Ropkins K. Openair—an R package for air quality data analysis. *Environ Model Softw.* 2012:27:52–61.

[108] Viglione A and Parajka J. TUWmodel: Lumped/Semi-Distributed Hydrological Model for Education Purposes. 2020. Available from: https://cran.r-project.org/package=TUWmodel

[109] Gudmundsson L, Bremnes JB, Haugen JE, and Engen-Skaugen T. Technical Note: Downscaling RCM precipitation to the station scale using statistical transformations – a comparison of methods. *Hydrol Earth Syst Sci.* 2012;16:3383–3390.

[110] Montaseri RR. WRSS: Water Resources System Simulator. 2019. Available from: https://cran.r-project.org/package=WRSS

[111] Busetto L and Ranghetti L. MODIStsp: an R package for preprocessing of MODIS land products time series. *Comput Geosci.* 2016;97:40–48. Available from: https://github.com/ropensci/MODIStsp

[112] Time Series Cran. Available from: https://cran.r-project.org/web/views/TimeSeries.html

[113] O'Hara-Wild M, Hyndman R, and Wang E. fable: Forecasting Models for Tidy Time Series. 2021. Available from: https://cran.r-project.org/package=fable

[114] Hyndman RJ and Khandakar Y. Automatic time series forecasting: the forecast package for R. *J Stat Softw.* 2008:29(27):1–22.

[115] Taylor S and Letham B. prophet: Automatic Forecasting Procedure. 2021. Available from: https://cran.r-project.org/package=prophet

[116] Trapletti A, Hornik K, and LeBaron B. tseries: Time Series Analysis and Computational Finance. 2021. Available from: https://cran.r-project.org/package=tseries

[117] Di Narzo AF, Aznarte JL, and Stigler M. tsDyn: Nonlinear Time Series Models with Regime Switching. 2020. Available from: https://cran.r-project.org/package=tsDyn

[118] Tsay RS, Wood D, and Lachmann J. MTS: All-Purpose Toolkit for Analyzing Multivariate Time Series (MTS) and Estimating Multivariate Volatility Models. 2021. Available from: https://cran.r-project.org/package=MTS

[119] Laurinec P. TSrepr R package: Time series representations. *J Open Source Softw*, 2018:3(23):577. Available from: https://doi.org/10.21105/joss.00577

[120] Aldrich E. wavelets: Functions for Computing Wavelet Filters, Wavelet Transforms and Multiresolution Analyses. 2020. Available from: https://cran.r-project.org/package=wavelets

[121] Machine Learning Cran. Available from: https://cran.r-project.org/web/views/MachineLearning.html

[122] Venables WN and Ripley BD. Modern Applied Statistics with S. 2002. Available from: https://www.stats.ox.ac.uk/pub/MASS4/

[123] Rong X. deepnet: Deep Learning Toolkit in R. 2014. Available from: https://cran.r-project.org/package=deepnet

[124] Therneau T, Atkinson B, and Ripley B. rpart: Recursive Partitioning and Regression Trees. 2019. Available from: https://cran.r-project.org/package=rpart

[125] Witten IH, Frank E, Hall MA, and Pal CJ. Practical machine learning tools and techniques. In *Data Mining* 2005, vol. 2, no. 4. Elsevier.

[126] Liaw A and Wiener M. Classification and regression by randomforest. *R. News*, 2002;2:18–22. Available from: https://cran.r-project.org/doc/Rnews/

[127] Chen T, Tong He T, Benesty M, *et al*. xgboost: Extreme Gradient Boosting. 2022. Available from: https://cran.r-project.org/package=xgboost

[128] Sekhon JS and Mebane WR. Genetic optimization using derivatives. *Polit Anal*. 1998;7:187–210.

[129] Riza LS, Bergmeir C, Herrera F, and Benítez JM. frbs: Fuzzy rule-based systems for classification and regression in R. *J Stat Softw*, 2015;65(6), 1–30. Available from: http://www.jstatsoft.org/v65/i06/

[130] Campbell JB and Wynne RH. *Introduction to Remote Sensing (5th ed.)*. New York, NY: Guilford Press, 2011, pp. 117–118.

[131] Strong JA and Elliott M. The value of remote sensing techniques in supporting effective extrapolation across multiple marine spatial scales. *Mar Pollut Bull*. 2017;116(1–2):405–419. Available from: http://dx.doi.org/10.1016/j.marpolbul.2017.01.028

[132] Schwalb-willmann J, Benjamin A, Horning N, Hijmans RJ, and Leutner MB. R: RStoolbox: Tools for Remote Sensing Data Analysis. 2018. Available from: https://cran.r-project.org/package=RStoolbox.

[133] Goslee SC. Analyzing remote sensing data in R: the landsat package. *J Stat Softw*, 2011;43(4):1–25. Available from: https://doi.org/10.18637/jss.v043.i04

[134] Lehnert LW, Meyer H, Obermeier WA, *et al*. Hyperspectral data analysis in R: the hsdar package. *J Stat Softw*. 2019;89:1–23.

[135] Lamigueiro OP and Hijmans R. rasterVis: Visualization Methods for Raster Data", 2023. Available from: https://oscarperpinan.github.io/rastervis/

Chapter 12

Geospatial big data analysis using neural networks

Hemi Patel[1], Jai Prakash Verma[1], Sapan Mankad[1], Sanjay Garg[2], Pankaj Bodani[3] and Ghansham Sangar[3]

Geospatial information system (GIS) produces large and complex data. Geo-spatial big data analysis is a critical field nowadays because a large amount of data is generated every day by various space mission programs running through space agencies all over the globe. It requires robust data storage and retrieval systems for decision-making in various GIS-based systems. This paper introduces a method for data analysis by adding a fog layer in the cloud-based GIS. The Fog environment is integrated mainly for data pre-processing and data cleaning for GIS systems. The load on the cloud environment will reduce when the pre-processing tasks are executed on the fog layer. The Weather dataset is used for weather prediction using an artificial neural network in a cloud environment.

12.1 Introduction

Big Data Analysis is a significant concern nowadays because the amount of data generated in a day is around 2.5 quintillion bytes [1]. Therefore, we need technologies that pre-process these data for various applications. Furthermore, we must develop a mechanism that handles extensive geospatial data for the data generated by satellites or geographic information system (GIS). This chapter discusses methods, tools, technologies, and platforms for handling big geospatial data efficiently. (i) Google BigQuery GIS is used for data generated from GIS and deals with a large dataset for finding the information. Therefore, it is the most powerful tool for significant dataset analysis from google. (ii) Open-source libraries and binaries like Python GDAL binding are used to process the data. First, it will divide the large data set into small chunks and find the results using libraries. Then it will merge all the results from these small chunks and generate the final result. (iii) SpatialHadoop –

[1]Institute of Technology, Nirma University, India
[2]CSE Department, Jaypee University of Engineering and Technology, India
[3]Space Applications Centre (SAC), Indian Space Research Organisation (ISRO), India

Apache Hadoop deals with big data analysis, and it has a spatial version for geospatial big data analysis, known as Spatial Hadoop. (iv) Google Earth Engine – this tool contains a large data set, and we can use and manipulate these real-time spatial data sets without downloading. (v) AWS Athena – this gives a simple interface to perform a query on storage using SQL. (vi) NoSQL and graph databases – databases like MongoDB and Elastic search also support some functionalities for geospatial data sets. Following are some terminologies related to geospatial data analysis.

12.1.1 Geospatial data

Geospatial data consists of objects, events, or attributes near the earth's surface. There are three types of data: Location data, Attribute data, and Temporal data. Location data is used to define the coordinates on the earth's surface. Location data can be static or dynamic. The example of static location data is the location of any building or maps of the road. Dynamic data is data that changes with time and a moving car is an example of dynamic data. Attribute data is characteristic of any object, event, or geographical feature. These data can include numerical or descriptive data. Temporal data is the event according to time with respect to location. The example of geospatial data consists of vectors, attributes, high-resolution images taken using satellites, architectural data of the buildings, census data, phone location data and social media data. This data can be vector data or raster data. Vector data is used for the representation of the buildings, houses or coordinates of roads, etc. Raster data is used for Image data which consists of pixels, and it is represented using rows and columns. Main concern for geospatial data is to collect and manage these data. Geospatial data can be used in agriculture, weather forecast, wildfire mapping, storm response, etc.

12.1.2 Big data analysis

Data analysis is divided into four parts. The first step is to collect or extract the data from the data source. After collecting, we have to clean the data and transform these data into a structural form, and the last step is to visualize these data through available platforms for extraction of meaningful information from the data. Data mining is the subprocess of Data analysis. In the statistical application, Data analysis is divided into the following parts: descriptive statistics, exploratory, and confirmatory data analysis [2,3].

12.1.3 Fog computing

Fog computing is a decentralized computing infrastructure that is added in between data sources and cloud infrastructure for increasing the computing power, storage, and networking services. The device used for providing above-mentioned services is known as the Fog node. In this chapter term, the fog layer was used, which represents the Fog environment in the overall infrastructure.

12.1.4 Neural network

A neural network in machine learning works the same as the human brain. It contains layers and neurons that are the same as the brain. It will make decisions based

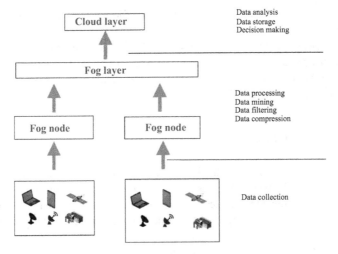

Figure 12.1   Fog environment

on mathematical functions defined for each neuron. It is known as the activation function. A neural network contains input, output, and hidden layers. It will help the model to define the output function. Each neuron is connected to another neuron and has a weight and bias assigned to it. At each layer, the neuron is activated and sends that data to the next layer; likewise, these processes repeat for all activated neurons, and the output is decided at last based on these values.

This chapter will discuss the big geospatial data analysis method using a Fog environment (refer to Figure 12.1). In this method, the load is divided into two parts: the first is on the Fog layer and another is on the cloud layer. Data analysis is performed on the Cloud layer [4], and Data mining or data pre-processing will be performed on the Fog layer. Using this method, we distribute the load using the Fog environment, and it will also reduce the load on the cloud environment for further processing [5].

This chapter contains six sections: the first section introduces geospatial big data analysis. The second section discusses the work done so far and different methods for geospatial big data analysis and strategies for data pre-processing. The third section contains the proposed work. The fourth section describes the methodology. The fifth section includes the experiment results, and the last section contains future work and a conclusion.

12.1.5   Contribution

This chapter addresses the future enhancement identified for the method mentioned in the paper City Geospatial Dashboard, IoT and Big Data Analytics for Geospatial Solution Providers in Disaster Management [6]. A dashboard is created for data visualization. It is used for data collection, data sharing, and data visualization. These data are collected from satellites, IoT devices, and other big data sources. The paper's objective is to improve the method by adding one layer to this system and sharing the load between these layers. The basic idea is to add a Fog layer and perform some data pre-processing to reduce the cloud layer's burden. Figure 12.1 contains three layers: the first layer is the data collection layer. In this layer, data is collected from different devices and sensors, and it sends all the collected data to the upper layer, known as Fog layer. Here, it will perform data pre-processing and data filtering. The basic definition is mentioned in the introduction. After applying these methods, data is now clean to perform data analysis on the cloud layer.

12.2   Related works

There are many methods and technologies for big data analysis on geospatial data of the existing methods, some methods are discussed below in Tables 12.1–12.4. The first subsection contains techniques for big data analysis on geospatial data, and the second subsection contains different data processing techniques for fog environment.

Table 12.1   FogGIS

Paper name	FogGIS: computing for geospatial big data analysis
Approach	FogGIS using Intel Edison and embedded micro processor
Objective	Develop fog computing-based data mining prototype
Methodology	Uses the fog computing, data mining and the following software, hardware and methodology is used. hardware: Intel Edison. Method: lossless decomposing technique
Pros	Fog gateway reduces the storage space requirements, transmission power, increase the throughput, reduce latency, and introduce the edge intelligence in geospatial cloud environment
Future work	More intelligent processing in fog layer
Data	Alaska, USA

Table 12.2   GeoBD2

Paper name	GeoBD2: geospatial big data deduplication scheme in fog-assisted cloud computing environment
Approach	GeoBD2: geospatial big data deduplication scheme
Objective	Derives the geospatial data deduplication structure for fog layer so that we can avoid the duplicate data
Methodology	Uses the height balance tree mechanism for deriving of Geo-DHBT structure
Pros	Minimum storage overhead cost than the existing big data deduplication scheme in cloud and fog environments
Cons	Better security mechanisms are required
Future work	The scheme will apply to mist computing systems for better management in cloud storage. It will also focus on the other encryption security mechanism for the improvement of secure and efficient geospatial data deduplication in the given computing environment

Table 12.3   Big data processing using fog computing

Paper name	Maximum data-resolution efficiency for fog-computing supported spatial big data processing in disaster scenarios
Approach	Spatial big data resolution using fog computing
Objective	In this paper, they perform spatial clustering for analysis and then integrate the fog layer for big spatial data analysis
Pros	Data is compressed at fog distributed layer
Cons	We can improve this algorithm by using different clustering algorithms
Future work	Introduction to data reduction ratio using a mathematical model, improve the quality of the clustering, a faster and more efficient algorithm for improvising the solution
Data	–

Table 12.4   GIS cloud computing

Paper name	GIS cloud computing-based government big data analysis platform
Approach	GIS – big data analysis in cloud computing
Objective	Analyse the data
Methodology	First analyses the spatial association analysis method of government big data, then proposes the architecture and function of GIS cloud computing-based government big data platform, and finally explores the application case based on traffic accident data
Pros	Business function customizer, report query customizer, workflow customizer
Cons	Need to select the attributes that may be analysed in the vector data as mining items
Data	GIS cloud-computing-based government big data analysis platform

12.2.1   Big data analysis on geospatial data

The first method is Hadoop and Map-reduce. HDFS and Hbase are used for data management, and map-reduce is used for splitting data and then running programs on all these data parts [4,7]. Hadoop is used for data processing, management, and storage [5]. Spark can be used for batch processing, streaming analytics, machine learning, graph databases, and ad hoc queries. Another tool is presto which is used for optimizing the SQL query. Hive is also used for SQL query processing, and the last is Hbase which is used as a database [8]. The second method is optimized Hadoop, in which all three phases of Hadoop are optimized, and the data is managed on it [9]. There is another method in which compression is performed on data [10,11]. The third type is GIS [12] and ArcGIS, which first analyses the spatial association analysis method of government big data. After that proposes the architecture and function of the GIS cloud computing-based government big data platform and finally explores the application case based on traffic accident data [13]. The fourth method for geospatial big data analysis is mist computing which will apply the algorithms for finding all the unused microcomputers in the computing environment [6]. The last big data analysis technique is cloud computing [14,15]. Tasks like data storage, data analysis, data pre-processing, data mining, and data management are performed on the cloud. Another modified version of cloud computing is hybrid cloud computing [16]. This architecture facilitates geospatial big data processing in hybrid cloud environments by leveraging and extending standards released by the Open Geospatial Consortium (OGC). Many more big data analysis techniques are used for geospatial data [17].

12.2.2   Data processing techniques in fog environment

This subsection contains the data mining techniques used in fog layer after integrating the fog layer into a cloud environment. The techniques are discussed below. The first technique is data deduplication in the fog layer. It will eliminate duplicate data from the data set and use the height-balanced tree mechanism to derive the Geo-DHBT structure [18]. It checks for the AVL tree property at every level and balances the tree by rotating the left, right, or both. The second method is pattern recognition and feature selection – the first task is to identify potential threat patterns on the incoming data streams from sensors using machine learning algorithms. The second one is to perform feature extraction for the computing. The third method is data mining with different stages like dynamic time warping, data compression, speech recognition, data and bandwidth reduction – data compression and processing steps. Another method is the clustering algorithm for data compression, which analyses the spatial clustering process, a specific category for spatial data analysis [19]. It proposes an architecture to integrate data processing into fog computing. The last method is the swarm Decision Table – it has three steps for making a table: (1) select efficient features, (2) select the efficient rules, and (3) construct the rule matrix. There are 13 types of swarm feature selection algorithms: Best First, Particle Swarm Optimization (PSO) algorithm, Ant Colony Optimization algorithm, Bat Search algorithm, Bee Search algorithm, Cuckoo Search algorithm, Elephant Search algorithm, Firefly Search algorithm, Flower Pollination algorithm, genetic algorithm (GA), Harmony Search algorithm, Wolf Search algorithm, and Evolutionary algorithm [5]. The next section introduces the proposed work as below.

12.3   Proposed work

As per Figure 12.2, this chapter proposed a method for load distribution on cloud layer and added one layer called the fog layer, which uses fog computing. The basic idea is to distribute the work like data pre-processing on this layer for the neural network-based data analysis model. Artificial neural network (ANN), Convolutional Neural Network (CNN), or Recurrent Neural Network (RNN) can be used for data analysis on the cloud layer. In this paper, ANN is discussed in detail and mentioned the drawback of ANN for image processing on the cloud layer. For data analysis, the first task is to select a model, and learning algorithm and decide the activation function. These steps are discussed in the next section in detail.

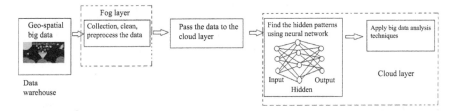

Figure 12.2   Block diagram

For weather prediction, ANN is used. A neural network contains three layers: input layer, output layer, and hidden layer. Each layer contains weights, bias, and activation functions. For training the data, there are three ways: first is supervised learning in which input and output are provided. The second method is unsupervised learning, in which the input is provided, and the model will find the hidden pattern. The last method is reinforcement learning; input and output are provided, and the task is to find the hidden pattern. NN is used for tabular data, image data, and text data. In this paper, the geo-spatial data set is used for training which consists of longitude, latitude, date, and time information.

12.4   Methodology and concepts

12.4.1   Data pre-processing on fog environment

As mentioned in the previous section, the load is reduced on the cloud layer by adding one layer called the fog layer. In this methodology, there are two options for reducing the load on the cloud layer: fog computing or edge computing [20]. But in this chapter, fog computing is used. The fog layer is used for data pre-processing, data cleaning. It will add missing values with the mean values, and as mentioned in Table 12.1, it takes 6–7 min for data pre-processing for nearly 1.8k records. Time will increase as we increase the number of data records. So by introducing the fog layer, time and processing power are reduced on the cloud. In this portion of the overall system, the data is collected from different sensors (data like pressure, temperature, wind speed, etc.) and then merged this data based on date and time to create a data frame for weather prediction based on data. Another task is to fill in the null or missing values in the data. These null values are filled by mean values using the inbuild function. This is the basic functionality of the fog layer in the system. For implementation, the local machine is used as the fog layer.

12.4.2   Prediction on cloud environment using ANN

As shown in Figure 12.3, this section briefly introduces the basic structure of the system. ANN consists of three layers: an input layer, an output layer, and a hidden layer, and it also contains weights and biases. For each neuron, there is an activation function used for switching on or off the functionality of the neuron. If the

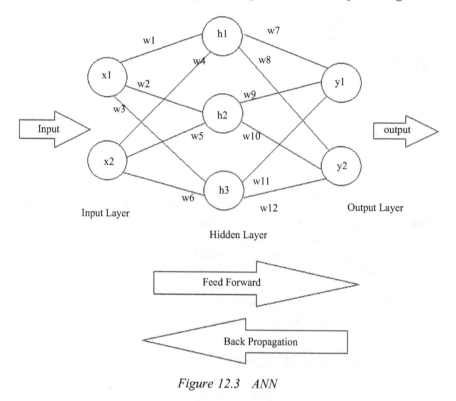

Figure 12.3   ANN

activation function is not used, it will only use a linear function for calculation, which is not a good idea. So, the activation function is like a powerhouse of the ANN. So, one neuron or perceptron is one logistic regression, and ANN is a group of neurons.

Here, weights are numeric values. The first step is to give the input, find the output value, and find the loss function based on predicted and actual values. This process is known as feed-forward. If the weights are updated of the ANN and repeat the process until getting good accuracy, it is known as backward propagation. But two basic problems occur in ANN: the first is the vanishing gradient and exploding gradient. A vanishing gradient occurs when the weight updating is less or does not change after some time or epoch. Exploding gradient occurs when the change value is large. The next section will discuss the implementation part and compare the results.

12.5   Results and discussion

Data set information: Weather prediction dataset selected from Kaggle (source: https://www.kaggle.com/alihanurumov/weather-prediction-network/data). The dataset contains seven files for temperature, pressure, wind direction, wind speed, city attribute, humidity and weather description. Pre-processed data and convert it into the proper data frame by removing null values (by putting mean values). Then apply ANN with three layers and the activation function used is sigmoid. The proposed model gets an accuracy of 70% for 1,800 records and 3,600 records.

Data pre-processing on fog environment: platform configuration: As shown in Table 12.5, the implementation is done on Google colab, local pc is considered as the fog layer. So, data pre-processing is done on the fog layer. As the results mention, almost 6 min are saved by pre-processing data on the fog layer for 1.8k records and 13 mins for 3.6k records. By using the fog layer, the CPU's power consumption and processing power are reduced on the cloud layer.

Prediction on cloud environment (comparison of algorithm on cloud layer – execution steps):

- Train/test split: 70% train and 30 % split
- Label encoding: one hot encoding
- Three layers:
  - Input layer:
    - Activation function: Sigmoid
    - Input features/independent variables: 5 (temperature, pressure, wind speed, wind direction, humidity)
    - Number of neurons and input shape: 5
    - Kernel initialization: he_normal

  - Hidden layer (ANN-1 layer):
    - Activation function: Sigmoid
    - Kernel initialization: he_normal
    - Number of neurons: 16

  - Output layer:
    - Activation function: Sigmoid
    - Kernel initialization: he_normal
    - Dependent variable: 1 (Weather description)

- Optimizer: adam
- Loss function: binary_crossentropy

As shown in Table 12.6, the algorithms have an accuracy of 70%. At first, all the algorithms were applied for nearly 1.8k records for different dates and times in

Table 12.5 Platform configuration

Implemented on	Google colab
RAM used	1.34 GB
Runtime for data preprocessing	6 min (for 1.8k records)13 min (for 3.6k records)
CPU model name	Intel(R) Xeon(R)
Available RAM	12 GB
Disk space	25 GB
CPU freq.	2.3 GHz

Table 12.6   Comparison of different
algorithms

Method	Accuracy (1,800 record)
ANN	0.67876588 (67.87%)
Decision tree	0.72776769 (72.77%)
KNN	0.722323049 (72.23%)
Gaussian NB	0.72776769 (72.77%)
SVM	0.72 (72%)

a particular area. The same process was applied for 3.6k records and got an accuracy of almost 71%. So, all the files contain the date, time, longitude, latitude, and weather prediction attributes (temperature, pressure, humidity, wind speed, and wind direction).

12.6   Conclusion

This paper discussed the fog layer and neural networks on the cloud layer. A fog layer is introduced in cloud computing to reduce the load on the cloud layer. On this layer, a task like data pre-processing is performed. A local machine is used as a fog layer and performs data pre-processing on the weather dataset. As shown above, a time of about 6 min is saved for 1.8k records. Time will increase as the number of records increases. Other than time, the processing power is reduced on the cloud layer. And for weather prediction, an ANN is used. Here, nearly 71% accuracy is achieved for 3.6k data records. The results are compared with algorithms like decision trees, KNN, Gaussian NB, and SVM. We can use CNN, RNN, etc., for better results.

References

[1]   C. Dobre and F. Xhafa, "Intelligent services for big data science," *Future Generation Computer Systems*, vol. 37, 2014, pp. 267–281, ISSN 0167-739X.

[2]   D. Li, Y. Gong, G. Tang, and Q. Huang, "Research and design of mineral resource management system based on big data," in *5th IEEE International Conference on Big Data Analytics*, 2020.

[3]   L. Geng1 and P. Xue, "Research on the protection and utilization evaluation of historic and cultural streets based on big data technology," in *2nd International Conference on Machine Learning, Big Data and Business Intelligence (MLBDBI)*, 2020.

[4]   S.K. Siledar, B. Deogaonkar, N. Panpatte, and J. Pagare, "Map reduce overview and functionality," in *Proceedings of the 6th International Conference on Communication and Electronics Systems*, 2021.

[5] D. Li, Y. Gong, G. Tang, and Q. Huang, "GeoBD2: geospatial big data deduplication scheme in fog assisted cloud computing environment," in *8th International Conference on Computing for Sustainable Global Development*, 2020.

[6] K.K., Sekimoto, Y., Takeuchi, W., *et al.*, "City geospatial dashboard: IoT and big data analytics for geospatial solutions provider in disaster management," in *2019 International Conference on Information and Communication Technologies for Disaster Management (ICT-DM)*. IEEE, 2019.

[7] P.R. Merla and Y. Liang, "Data analysis using Hadoop MapReduce environment," in *IEEE International Conference on Big Data (BIGDATA)*, 2017.

[8] M. Dick, J.G. Ji, and Y. Kwon, "Practical difficulties and anomalies in running Hadoop," in *International Conference on Computational Science and Computational Intelligence*, 2017.

[9] V. Sontakke and R.B. Dayanand, "Optimization of Hadoop MapReduce model in cloud computing environment," in *Second International Conference on Smart Systems and Inventive Technology*, 2019.

[10] P. Praveen, C.J. Babu, and B. Rama, "Big data environment for geospatial data analysis," in *2016 International Conference on Communication and Electronics Systems (ICCES)*. IEEE, 2016.

[11] K. Rattanaopas and S. Kaewkeeree, "Improving hadoop mapreduce performance with data compression: a study using wordcount job," in *14th International Conference on Electrical Engineering/Electronics, Computer, Telecommunications*, 2017.

[12] D. Sik, K. Csorba, and P. Ekler, "Implementation of a geographic information system with big data environment on common data model," in *IEEE*, 2017.

[13] Krämer, M. and I. Senner, "A modular software architecture for processing of big geospatial data in the cloud," *Computers & Graphics*, vol. 49, 2015, pp. 69–81.

[14] Q. Wang and Y. Jiang, "GIS cloud computing based government Big Data analysis platform," in *IEEE 2nd International Conference on Big Data, Artificial Intelligence and Internet of Things Engineering*, 2021.

[15] Huang, Y., P. Gao, Y. Zhang, and J. Zhang, "A cloud computing solution for big imagery data analytics," in *2018 International Workshop on Big Geospatial Data and Data Science (BGDDS)*, IEEE, 2018, pp. 1–4.

[16] S. Ingo, "Geospatial Big Data processing in hybrid cloud environments," in *IGARSS 2018–2018 IEEE International Geoscience and Remote Sensing Symposium*, IEEE, 2018, pp. 419–421.

[17] R.K. Barik, S.S. Patra, P. Kumari, S.N. Mohanty, and A.A. Hamad, "New energy-aware task consolidation scheme for geospatial big data application in mist computing environment," in *IEEE*, 2019.

[18] H. Sulimani, W.Y. Alghamdi, T. Jan, G. Bharathy, and M. Prasad, "Sustainability of load balancing techniques in fog computing environment: review," *Procedia Computer Science*, vol. 191, 2021, pp. 93–101, ISSN 1877-0509.

[19] Md. A. Elaziz, L. Abualigah, and I. Attiya, "Advanced optimization technique for scheduling IoT tasks in cloud-fog computing environments," *Future Generation Computer Systems*, vol. 124, 2021, pp. 142–154, ISSN 0167-739X.

[20] R.K. Barik, H. Dubey, A.B. Samaddar, *et al.*, "FogGIS: fog computing for geospatial big data analytics," in *2016 in IEEE Uttar Pradesh Section International Conference on Electrical, Computer and Electronics Engineering (UPCON)*. IEEE, 2016.

Chapter 13

Software framework for spatiotemporal data analysis and mining of earth observation data

KP Agrawal[1], Pruthvish Rajput[2], Shashikant Sharma[3] and Ruchi Sharma[4]

Spatiotemporal data are generated in various fields: such as agriculture, defence, meteorology, crop sciences, medicine, and transportation. Data can be captured in various formats at multiple levels of granularity both in space and time. Such data can be analyzed using various visualization methods, spatiotemporal On Line Analytic Process (OLAP) operations and data mining. Separate tools exist for each possible way of analysis, which lacks in some other way in fulfilling generic functionalities to analyze each sort of spatiotemporal data. The paper proposes a framework, spatiotemporal data analysis and mining environment (ST-DAME) for effective analysis of spatiotemporal data by integrating various components of geographic visualization, information visualization, OLAP server, and data mining server in a single software. Machine learning and deep learning techniques can be used for clustering purposes using spatiotemporal data to expedite the work faster. Moreover, other than a generalized framework, the need for application-specific analysis is also realized and incorporated.

13.1 Introduction

Space and time are the basic dimensions of our existence and thus needed to explore and analyze to reveal hidden knowledge in various fields, from the micro level of cells in our physical body to the macro level of our planet Earth and its geography.

In the past few years, many software tools have been developed for spatio-temporal data analysis. Many of them are application specific, which means they are developed and designed to be meant for analysis in the specific domain with limited functionalities. Some of them are only meant for spatial analysis; some are performing well enough to answer spatiotemporal queries but are incapable to resolved complex queries.

[1]Symbiosis Institute of Technology, Symbiosis International University, India
[2]U.V. Patel College of Engineering, Ganpat University, India
[3]Space Application Centre, ISRO, India
[4]Institute of Technology, Nirma University, India

Owing to such shortcomings of existing tools, authors have been prompted to incorporate features like 'firing MDX queries' at multiple granularities especially keeping in view spatiotemporal data and getting results in either automated or customized way depending upon the end-user requirements.

Tools developed till now to analyze spatiotemporal data use one of the following approaches.

13.1.1   Visualization

It is easy for a human being to interpret things by seeing them visually (as there is a saying a picture is worth a thousand words). The challenge, in building such tools, is to develop effective visualization components, using which users can interpret hidden information from the spatiotemporal data. The visual analysis feature has evolved to interpret data through visualization.

13.1.2   Multidimensional analysis

Many of the spatiotemporal queries can be answered by applying OLAP Cube operations such as Roll up, Drill down, Slicing, and Dicing. Some complex queries can also be answered via advanced data cube techniques like the following:

(i)   Discovery-driven data cubes
(ii) Multi-feature cubes: these can perform complex aggregations at multiple granularities

Performance considerations for operations like visualization of aggregated spatiotemporal data are required to be taken care of when developing such systems.

13.1.3   Data mining

Through online analytical processing we can get to know what is happening but very difficult to predict what will happen in the future and why it is happening.

For getting answers to such queries, we need to apply machine learning or deep learning techniques such as clustering, classification, association rule mining, and predictive analysis.

Here the focus of the authors is to get good quality clusters (micro and nested clusters) in a minimum time, for this purpose, machine learning and deep learning techniques (neural network with multiple hidden layers) can be incorporated.

Alvares *et al.* [1] propose various versions of Weka i.e. Weka-GDPM and Weka-STPM data mining framework for supporting spatiotemporal databases. The mentioned versions lack spatiotemporal data mining techniques, components for geo-visualization, and multidimensional analysis.

To effectively and efficiently analyze spatiotemporal data, the framework should offer the integration of all three approaches in a single workbench.

13.2   Related work

Several research works have been done in the area over the past few years. Tools for spatiotemporal data analysis vary from simple query environments to sophisticated data mining.

Spatiotemporal data mining is an emerging area of research. Applications for mining different kinds of spatiotemporal patterns and trends are being developed by researchers in various domains. However, there is a need for an underlying architecture framework for these applications to provide reusability of analysis and design [2–4]. Patel and Garg [5] discuss some issues and challenges in designing data mining architecture. Andrienko [6] discussed issues related to space, time and visual analytics. Patel and Garg [5] have also discussed the advantages of having distributed data mining architecture, which can be beneficial in the case of spatiotemporal data also due to the large size of such data. Cavalcanti *et al.* [7] present a visual query system which supports querying spatiotemporal databases. Such systems can answer queries simply by retrieving databases and presenting retrieved data to users in the form of maps and graphs.

For analyzing data in greater depth, some sort of processing over the stored data and also effective visualization are needed to be done. Processing over data may include multidimensional analysis and data mining. Shekhar *et al.* [8] propose the concept of a mapping cube and in 2002 presents its use in a specific application, named 'CubeView' [9] for analyzing and observing rapid summarization of major trends over traffic data. The application promotes the use of multidimensional analysis but at the same time is domain specific.

A system for image analysis integrates image analysis tools, metrics based on landscape ecology theory, multi-temporal feature handling, and data mining techniques [10]. Oliveira *et al.* [11] implemented an approach to analyze spatiotemporal data through visualization and clustering. Again, it is not complete in terms of analysis, as multidimensional data analysis is not possible.

Each system implements some sort of analysis approach or combination of them, to reveal useful information from spatiotemporal data. From the literature survey, we have realized that no tool till now provides full-fledged functionalities to analyze heterogeneous spatiotemporal data in every possible way.

13.3 Challenges

There are several issues with the existing spatiotemporal data analysis software that prompted us to think of designing and developing the proposed framework. They are as follows:

- Existing open-source GIS tools do not provide sophisticated data mining facilities.
- Multi-level granularities in space and time should be handled effectively.
- Many tools lack effective visualization components.
- Analysis software should be able to support spatiotemporal data in various possible formats.
- Software should provide all the possible functionalities for analysing spatiotemporal data inside a single workbench.

13.4   The ST-DAME

The framework should exhibit the following characteristics:

- Domain independence – not meant for specific applications only, will provide all possible functionalities needed for spatiotemporal data analysis in general.
- Customization – the system would provide full customization to the user i.e. selection of appropriate data format, sequence of tasks to be performed, and visualization all would be dependent on the user's action.
- Usability – easy to understand and use.
- Support for heterogeneous data – would support spatiotemporal data in various possible file formats and database systems.

13.4.1   Conceptual architecture of the framework

Major features, functionalities or components of the framework can be thought of conceptually, under three layers, as shown in Figure 13.1, for ease of development of the software.

13.4.1.1   Data layer

This layer is concerned with heterogeneous data types and database support. Spatiotemporal data can be found in any of the possible file formats and also in any of the database systems with spatial data support.

13.4.1.2   Functional layer

This layer represents all the processing tasks that can be applied to the data and are offered by the system. It would consist of three components for our framework: (i) pre-processing, (ii) data mining, and (iii) ST-OLAP operations.

13.4.1.3   Visualization layer

All the visualization resources offered by the framework come under this layer. Visualization is needed at each step of the system, such as: a user interface, input data, metadata, processed data, or information. The layer is one of the major

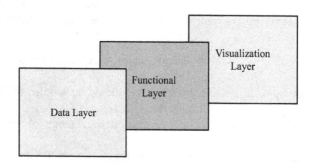

Figure 13.1   Conceptual architecture of the system

aspects of the overall system, as spatiotemporal data itself contains a visual dimension called space.

13.4.2 Proposed framework

The framework offers functionalities as shown in Figure 13.2.

The proposed framework has the following modules:

1. Input module
2. Pre-processing module
3. Data mining module
4. Multidimensional analysis module
5. Application generation module

The framework consists of the above modules, for accomplishing the following specific task or functionality:

13.4.2.1 Input module

It would be responsible for taking input in various forms and will exhibit characteristic for the support of heterogeneous data shown in Figure 13.3.

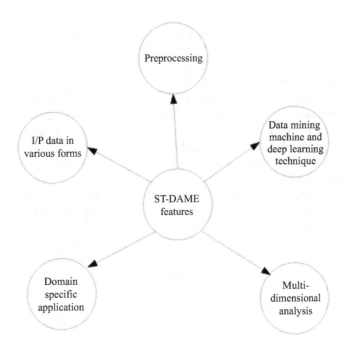

Figure 13.2 ST-DAME features

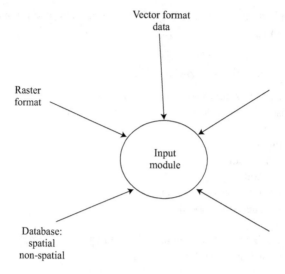

Figure 13.3   Various data formats supported by the input module

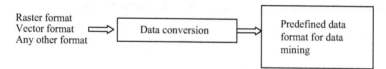

Figure 13.4   Data conversion before application of data mining technique

13.4.2.2   Pre-processing module

Spatiotemporal data is needed to be cleaned, transformed, and modified as per its usage. There are several pre-processing tasks such as scaling, matching spatial or temporal granularities of various data to be integrated, noise removal, etc. All such tasks would be made available through this module. Moreover, the task of data conversion would also be facilitated which is essential before the application of any data mining technique.

Since the data can be taken in various formats through the input module, it is needed to be converted into a predefined format, before the application of the data mining technique, as illustrated in Figure 13.4.

13.4.2.3   Data mining module

The spatiotemporal dataset exhibits some distinct properties, for which conventional data mining techniques may not perform well or may not be able to give appropriate results. These datasets are continuous whereas conventional datasets are mostly discrete in nature. They exhibit local patterns and conventional data mining techniques are developed for handling global patterns. Moreover, spatiotemporal data sets tend to be highly correlated. Autocorrelation among data must be

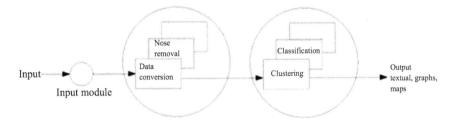

Figure 13.5   Data flow diagram for data mining

considered while applying data mining on such data. For the above-mentioned reasons, separate data mining techniques are needed for spatiotemporal data.

The data is needed to be processed through several modules as per the requirement of the user. Typical data flow for data mining is illustrated in Figure 13.5.

The data mining module should offer the features mentioned below:

- Integration of new techniques: it should provide the facility to add new techniques to the workbench or modify existing techniques.
- Application of existing techniques: all the techniques would take input data in the predefined format only, to fulfil this requirement data conversion is done to change the input format into the acceptable format by data mining technique(s).

13.4.2.4   Multidimensional analysis module

For analyzing spatiotemporal data over multiple dimensions and to summarize data at multiple levels of spatial and temporal granularities the framework facilitates interface to the user via this module. The data flow for the multidimensional analysis is illustrated in Figure 13.6. The user selects appropriate dimensions for making an MDX query graphically; the integrated spatiotemporal-OLAP server will accept those queries and convert them to SQL queries for retrieving content from a physical database.

The output will be displayed in the form of histograms, bar charts, and geographic maps.

13.4.2.5   Application generation module

Various modules of the framework are therefore sophisticated analysis and mining of spatiotemporal data.

This sort of analysis is meant for data mining analysts and technical users. For non-technical users, some easy-to-use user interface should be there, where users can fire queries by selecting a few appropriate times, space, and measure parameters and appropriate results would be displayed in the form of maps, graphs, or reports. The input–output process for domain-specific applications is illustrated in Figure 13.7. The module would provide the facility to generate new applications for a specific domain. That could be then launched as a stand-alone application or web application.

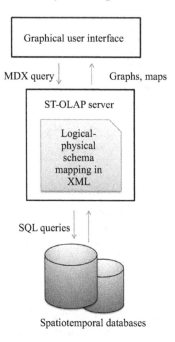

Figure 13.6   Data flow diagram for multidimensional analysis

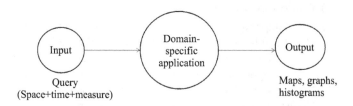

Figure 13.7   Input–output diagram of domain-specific application

The generated application would then be available to a mass of users to easily navigate through summarized information at multiple levels of granularity in space and time and to visualize the data in the form of maps and graphs.

13.4.3   ST-DAME in action

The integrated system can be used in several ways depending on the need. Depending on the input, two cases are as follows:

(i)   Dealing with flat files: when input data is a flat file whether raster, vector or textual; there is only one possible sequence of tasks, as shown in Figure 13.8:

(ii)   Dealing with data-warehouses spatiotemporal: datawarehouses can be handled and analyzed in different ways, using ST-DAME.

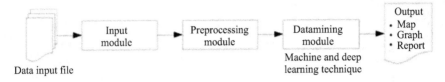

Figure 13.8 Dealing with flat files

Figure 13.9 Multidimensional analysis

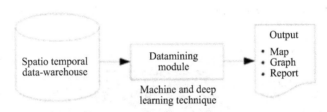

Figure 13.10 Spatiotemporal data mining

The possible ways are:

(a) Multidimensional analysis: the user can analyze or summarize various measures across multiple dimensions and then can get output in various possible forms as shown in Figure 13.9.

(b) Spatiotemporal data mining: data mining techniques like clustering, classification and association rule mining may be applied to the data warehouse for getting desired output as shown in Figure 13.10. To expedite the work faster different machine learning techniques where learning to a neural network can be imparted or deep learning techniques with multiple hidden layers (to automate the process of feature extraction).

(c) Multi-dimensional analysis over-extracted information: the user would be able to use data mining and multidimensional analysis in combination. For example, a user may want multidimensional analysis over clustered data. This way is mentioned in Figure 13.11.

(d) Data mining over summarized data at multiple levels after application of multi-dimensional analysis for summarization at a desired level of granularity, the user can apply data mining technique on that level of abstraction as shown in Figure 13.12.

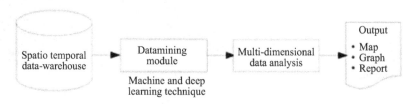

Figure 13.11   Multi-dimensional analysis of over-extracted information

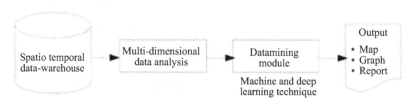

Figure 13.12   Data mining over summarized data at multiple levels

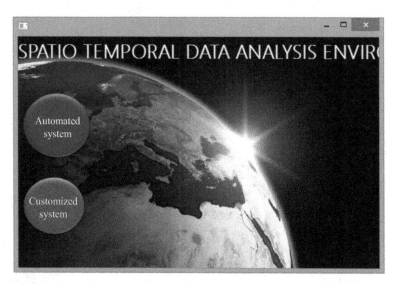

Figure 13.13   Main window for STDM software framework

13.5   Result

In our proposed system, we have designed and developed two separate systems for analysis which are named as:

1. Automated system
2. Customized system

At first, the start window as shown in Figure 13.13 is splashed over the screen, asking for an option to be selected between the two. Depending on the choice of the user, a user interface for the selected subsystem is displayed next.

13.5.1   *Automated system*

An automated system is meant for multi-dimensional analysis and visualization, where a data warehouse has been created. Input to the automated system is a combination of spatial, temporal, and non-spatial dimensions. The spatial dimension consists of three levels of granularities i.e., country, state and district and the temporal dimension have three levels of granularities i.e. year, quarter, and month. Depending on the need of end users, a multi-dimensional query would be formed automatically which will be taken care of by the underlying system developed. The visual interface offers three groups of selection controls: measures, space dimension, and time dimension.

> Group 1: Measures offer parameters to select from. In our selected domain, these are vegetation index, rainfall, and temperature. Measures are dependent on the area of the domain and should be specified accordingly in the interface.
> Group 2: Time dimension has three levels of granularity i.e., year, season (i.e., Kharif, Rabi, and Zaid), and month. A further selection of particular time will be displayed up to a selected level of granularity.
> Group 3: Space dimension also offers three levels of granularity to select from, i.e., country, state, and district. Choices displayed for the lower level will depend on the higher level selected.

Results are displayed based on measures and selected level of granularity for space and time. For Figure 13.14, selection measures are as follows:

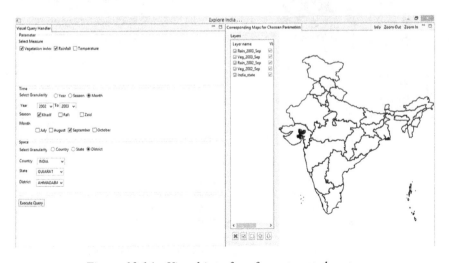

Figure 13.14   Visual interface for automated system

Group 1: Vegetation Index and Rainfall measures.

Group 2: (i.e., time dimension): year from 2002 to 2003, Kharif season and September month.

Group 3: (i.e., space dimension): Ahmedabad city of Gujarat, India.

Visualization of query results

Output is displayed in the form of maps in the cascaded panel as shown in Figure 13.14 and the status of NDVI and rainfall for selected spatial and temporal granularities is shown in Figure 13.15. Moreover, it also gives the exact value for a particular measure (NDVI/rainfall) by placing the cursor on point of interest.

Map windows can be zoomed in/out and layers of a map can be made visible or invisible according to the user's need as shown in Figure 13.16. The toolbar of the map panel facilitates some functionalities i.e., zoom-in, zoom-out, and information display at the selected point of location.

13.5.2   Customized system

Customized system takes care of data mining-related tasks manually depending on the need of end users. To perform these tasks, a tool developed by us has been incorporated in an existing open-source data mining tool called WEKA [12] to effectively utilize time for incorporating required functionality specially on the spatiotemporal database.

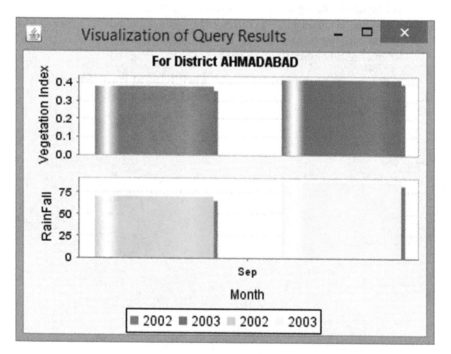

Figure 13.15   NDVI and rainfall status for selected spatial and temporal granularities

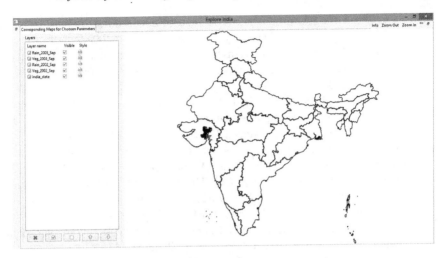

Figure 13.16   Map panel and layers utility

13.6   Conclusion

The proposed framework has offered generic functionalities by integrating three approaches for spatiotemporal data analysis that are data mining, multi-dimensional analysis (MDX queries) and visualization i.e., generation of graphs and maps after a rigorous survey of existing tools that are meant for such analysis. This resulted in overcoming issues of existing tools mentioned earlier and ease of analysis of spatiotemporal data. The framework also proposes an automated and customized spatiotemporal multidimensional data analysis environment where depending on end-user requirements, the result of MDX queries can be obtained which is evident from the result section. To deal with a huge amount of data, deep learning techniques (CNN/LSTM/RNN) can be used in future.

References

[1]   Alvares LO, Palma A, Oliveira G, *et al*. Weka-STPM: from trajectory samples to semantic trajectories. In: *Proceedings of the XI Workshop de Software Livre, WSL*. Citeseer, 2010, vol. 10, pp. 164–169.

[2]   Rao KV, Govardhan A, and Rao KC. An architecture framework for spatiotemporal datamining system. *International Journal of Software Engineering & Applications*. 2012;3(5):125.

[3]   Karssenberg D, Schmitz O, Salamon P, *et al*. A software framework for construction of process-based stochastic spatio-temporal models and data assimilation. *Environmental Modelling & Software*. 2010;25(4):489–502.

[4]   Andrienko N and Andrienko G. A visual analytics framework for spatio-temporal analysis and modelling. *Data Mining and Knowledge Discovery*. 2013;27(1):55–83.

[5]   Patel S and Garg S. Issues and challenges in data mining architecture. In: *National Conference on Emerging Technology in C.S.*, 2007.

[6]   Andrienko G, Andrienko N, Demsar U, *et al.* Space, time and visual analytics. *International Journal of Geographical Information Science*. 2010;24 (10):1577–1600.

[7]   Cavalcanti VM, Schiel U, and de Souza Baptista C. Querying spatio-temporal databases using a visual environment. In: *Proceedings of the Working Conference on Advanced Visual Interfaces*; 2006, pp. 412–419.

[8]   Shekhar S, Lu C, Tan X, *et al.* A visualization tool for spatial data warehouses. *Geographic Data Mining and Knowledge Discovery*. 2001;73:16–72.

[9] . Shekhar S, Lu CT, Liu R, *et al.* CubeView: a system for traffic data visualization. In: *Proceedings. The IEEE 5th International Conference on Intelligent Transportation Systems*. IEEE, 2002, pp. 674–678.

[10] Körting TS, Fonseca LMG, and Câmara G. GeoDMA—geographic data mining analyst. *Computers & Geosciences*. 2013;57:133–145.

[11] de Oliveira MG and de Souza Baptista C. An approach to visualization and clustering-based analysis on spatiotemporal data. *Journal of Information and Data Management*. 2013;4(2):134–134.

[12] Eibe F, Hall MA, and Witten IH. The WEKA workbench. In: *Online Appendix for Data Mining: Practical Machine Learning Tools and Techniques*. Morgan Kaufmann, 2016.

Chapter 14

Conclusion

Sanjay Garg[1], Kimee Joshi[2] and Nebu Varghese[3]

This book covers a wide range of topics about applications for Earth Observation data (EO) i.e. variety in applications, variety of data formats used, and variety of machine learning (ML) tools and techniques. This text covers applications for weather forecasting, crop monitoring, crop classification, LULC classification, hurricane, climate change, reservoir characterization, etc. Studies of various types of data such as optical data, metrological data, microwave data, spectral data, and hyperspectral data are considered for these applications. Various ML techniques like ANN, CNN, autoencoders, and auto-regression are used and also chapter on the R tool is included for comprehensive knowledge of the relevant tool.

14.1   Excerpts from various chapters

This book is divided into three parts. Chapters 2–6 deal with the clustering and classification of Earth Observation data, Chapters 7–10 deal with rare event detection in Earth Observation data, and Chapters 11–13 deal with tools and technologies for earth observation data.

Chapter 1 introduces various kinds of EO data sources with their characteristics and also various application areas for data analysis. A brief survey of various ML/DL/analytics techniques is also presented.

Chapter 2 considers the dimensionality reduction issue for hyperspectral remote sensing data for crop classification problems using the CNN model since each crop has a prominent signature because the hyperspectral data set has a greater number of bands. In this research, healthy and diseased crops can monitor using REP and can be classified using a deep learning method and concluded a better accuracy.

Chapter 3 presents the study of full polarization SAR data for crops and a deep learning-based crop classification method was used. Two methods Inception v3 and Custom VGG-like model are used by extracting fields of SAR images by super-

[1]CSE Department, Jaypee University of Engineering and Technology, India
[2]Institute of Technology, Nirma University, India
[3]Dholera Industrial City Development Limited, India

imposing the edge detected optical/SAR image on the Freeman decomposition image and labeled extracted SAR field images for crop classification. Inception v3 has demonstrated better performance than the Custom VGG-like model.

Chapter 4 contemplates a novel implementation and study for possible advancements in the field of LULC classification using DiceNets, DenseNets, ResNets, and SqueezeNets on the Indian Pines dataset. An embedding augmentation strategy called E-Mixup is also explored considering improvement to the standard or vanilla Siamese implementations.

Chapter 5 proposed a new remote sensing image classification algorithm using the clonal selection algorithm which is the basis of the immune system. The proposed approach is better capable of discriminating roof types and urban features than the conventional maximum-likelihood approach. It is a good and efficient classification algorithm and can be applied to remote sensing image classification.

Chapter 6 implemented supervised ML algorithms, a state-of-the-art SSD MobileNet-v2 for object detection, and U-Net for semantic segmentation and automatic detection of airplanes in UHSR images captured by UAV. The implemented architectures possess a limitation that objects that are similar in shape to the target i.e. airplane, are detected or segmented as targets too. The architecture for the segmentation of images performs hard(binary) classification for each pixel.

Chapter 7 worked on a transfer learning-based approach where various pre-trained CNN-based algorithms can be applied directly to enable recovery and rescue efforts after the landfall of a hurricane using geological data. An ensemble approach by training multiple pre-trained algorithms is used to increase the model's accuracy and utilized geological data to enhance the study of building damages.

Chapter 8 applied combinations of deep learning methods to satellite images for monitoring climate change and climate change has become a majorly discussed issue around the globe. This approach has also been proven to be more accurate and more reliable and faster than conventional methods.

Chapter 9 highlights plans to understand the viewpoints that impact avalanches in India's meteorological conditions. It discussed the course of avalanche occasions since planning avalanches is the initial step. It demonstrates that AI calculations and remote detecting information can propel this work and give an authentic contribution as a powerful occasion or catastrophe caution.

Chapter 10 studies reservoir characterisation, with an ML approach. Sensors data (seismic and well logs) are integrated to generate underlying porosity distribution of a considered prospect area using different ML techniques. This chapter provides good insight into the challenge, prospects, and potential research directions in this field of reservoir characterization.

Chapter 11 is a state-of-the-art analysis of the role of the R tool in remote sensing, hydrology, and hydrological research, focusing on the operational practice of hydrology by facilitating a wide range of hydrological analyses. Hydrology-related packages in R software are studied in detail and this is a good piece of text for the research community to understand the practical aspects of the tool.

Chapter 12 discusses the fog layer and neural networks on the cloud layer. The processing power is reduced on the cloud layer. A fog layer is introduced in cloud computing to reduce the load on the cloud layer and for weather prediction, an artificial neural network is used and finally accuracy of weather prediction is increased.

Chapter 13 fabricated a framework to offer generic functionalities by integrating three approaches for spatiotemporal data analysis that are data mining, multi-dimensional analysis (MDX Queries), and visualization. This resulted in overcoming issues of existing tools mentioned earlier and ease of analysis of spatiotemporal data. The framework also encompasses an automated and customized spatiotemporal multidimensional data analysis environment.

14.2   Issues and challenges

Deploying ML models for earth observation data has the many challenges, as illustrated in Figure 14.1. These challenges include:

(a) Massive volume data was collected from heterogenous sources e.g. optical, hyperspectral, microwave, metrological data, etc. and with different resolutions too.
(b) Reliability of data i.e. lack of ground truth data, which is major hurdle for building an accurate supervised ML model.
(c) Noise in EO data i.e. quality of data is a big challenge for deploying an ML model for EO data analysis.

14.2.1   Collecting meaningful and real-time data

Many scientific EO data sets, such as those from Landsat and Copernicus, have open access policies that allow for analyses on a continental or even global scale. The two types of satellite imagery that are most frequently used for earth

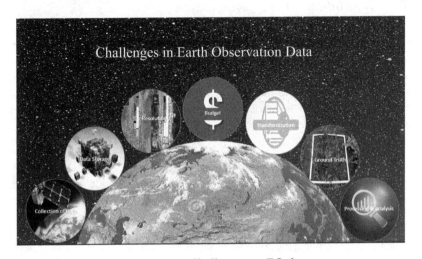

Figure 14.1   Challenges in EO data

observation are optical and radar images [1]. Understanding the characteristics and features of satellite imagery is essential to know how to use it. The requirements for data derived from satellite imagery can vary depending on a particular application [2]. The level of resolution, both temporal and spatial, is increasing as EO is applied to more local requirements. Aside from cost and computing requirements, the suitability of imagery is being questioned. This is especially true in humid areas where long-term cloud cover or critical periods (e.g., flooding) make optical imagery difficult to use [3]. For measuring crop health and doing vegetation analysis, the multispectral bands are commonly used [4]. The monitoring of the weather, agriculture harvest monitoring, field segmentation of icebergs, and natural or man-made disasters like tsunamis and oil spills are just a few of the applications of SAR satellite data [2].

14.2.2   Data storage

The amount of data that EO satellites are gathering keeps growing [5]. The traditional method of downloading image data to personal storage devices and performing the tasks on individual infrastructure and software is time-consuming and inefficient when dealing with large amounts of remote sensing data. Data is becoming increasingly "immobile" in the EO domain (e.g., data downloads have become inconvenient or impossible), necessitating the use of appropriate online tools to access and processing within the data environment. Because of technological advances many advancements in digital infrastructure, increased computing power and storage capabilities [1]. The phrase "EO data cu" (also known as "geospatial data cu" or "data cube onl" in some cases) has recently come to refer to a new method for organizing, managing, and analyzing EO data [1,6–9].

14.2.3   Resolution; quality promotion

The required resolution will depend on how big your project area may be and which features need to be seen properly. More different photos would be needed to cover such an area because better-resolution sensors often capture smaller area sizes. If complete coverage is present, it will likely be made up of photographs that have been gathered over several months or years. For a region of this size, lower-resolution alternatives will provide a better overview and typically require fewer photographs (and image dates) to fulfill the coverage [4].

14.2.4   Budget limitations

High-resolution imagery, which can be expensive, may provide an excellent analytical product for the time being, but the costs of replicating this for continuous monitoring may exceed available budgets. Furthermore, some satellite missions have finite lifetimes and may never be replaced [3]. For a research organization or a company that does not have an on-site laboratory, questions arise regarding the security of the data and algorithms developed [1].

14.2.5 Standardization

It is clear from the preceding that global data sets from EO are generated by a variety of sensors, processed by numerous organizations, and available from various sources. This means that the data have different resolutions and formats, so integration is required to make the best use of the data, and standardization would help with this process [5].

14.2.6 Lack of ground truth data

Remote sensing without ground truth data is constrained in the actual world by the complexity of earth's surface characteristics, the effects of the atmosphere, and the blurriness of spectral fingerprints. The synoptic picture supplied by satellites is complemented by the acquisition of ground truth data, which helps to relate the image data to the context of features of the earth's surface that are present on the ground. To comprehend images, ground truth is crucial [10]. To better understand data, ML requires labels, but the diversity of nature limits the application of ML algorithms. The existing categorization is frequently found to be insufficient for labeling data [11].

14.2.7 Processing and analysis

User-friendly interfaces are necessary for the context of a sustainable Earth where decision-makers who may not be able to make the best use of image-processing software packages need information. The information must be presented in a way that is understandable and easily manipulatable [5]. Many programs have shifted to free open-source software, although this takes greater degrees of knowledge to offer the features that private software does [3]. When asked about EO constraints associated with radar images and high-resolution optical data, more than half of the respondents mentioned difficulties in processing and interpreting radar data. One issue is the end-ability of user's to use radar images. Inadequate understanding of how to use radar images. Consider the Amazon region's dense cloud cover. It would be useful, but we would need proper training to process this type of data. More than 50% of respondents stated that the issues relate to both the technical abilities to handle EO radar data and the availability of radar images. One of the key needs identified by respondents was for current, high-resolution EO data with a resolution of 1m or less, and it was suggested that such high-resolution imagery should be freely accessible to the general public [12].

References

[1] Sudmanns M, Tiede D, Lang S, *et al*. Big Earth data: disruptive changes in Earth observation data management and analysis? *International Journal of Digital Earth*. 2020;13(7):832–850.

[2] ICEYE. Satellite Data: How to use Satellite Data for Better Decision Making. Available from: https://www.iceye.com/satellite-data.

[3] Addressing the Challenges of using Earth Observation Data for SDG Attainment: Evidence from the Global South and West Africa Region. Available from: https://static1.squarespace.com/static/5b4f63e14eddec 374f416232/t/624334541bf6aa586f04d166/1648571478139/Earth+Observation-3.29.pdf.

[4] Murray R. Choosing the right type of imagery for your project. 2021. Available from: https://www.l3harrisgeospatial.com/Learn/Blogs/Blog-Details/ArtMID/10198/ArticleID/15204/Choosing-the-right-type-of-imagery-for-your-project.

[5] Dowman I and Reuter HI. Global geospatial data from Earth observation: status and issues. *International Journal of Digital Earth*. 2017;10(4):328–341.

[6] Haklay M, Mazumdar S, and Wardlaw J. *Citizen Science for Observing and Understanding the Earth*. Springer; 2018.

[7] Baumann P, Rossi AP, Bell B, *et al.* Fostering cross-disciplinary earth science through datacube analytics. In: *Earth Observation Open Science and Innovation*. Springer, Cham; 2018. p. 91–119.

[8] Giuliani G, Chatenoux B, De Bono A, *et al.* Building an earth observations data cube: lessons learned from the swiss data cube (sdc) on generating analysis ready data (ard). *Big Earth Data*. 2017;1(1–2):100–117.

[9] Purss MB, Lewis A, Oliver S, *et al.* Unlocking the Australian landsat archive – from dark data to high performance data infrastructures. *GeoResJ*. 2015;6:135–140.

[10] Ground Truth Data Collection – egyankosh.ac.in. Available from: https://egyankosh.ac.in/bitstream/123456789/39538/1/Unit-9.pdf.

[11] Agarwal C. Challenges of using machine learning on Earth Observation Data; *Geospatial World*, 2021. Available from: https://www.geospatial-world.net/top-stories/challenges-of-using-machine-learning-on-earth-observation-data/.

[12] Cerbaro M, Morse S, Murphy R, *et al.* Challenges in using Earth Observation (EO) data to support environmental management in Brazil. *Sustainability*. 2020;12(24):10411.

Index

Printed in the USA
CPSIA information can be obtained
at www.ICGtesting.com
JSHW01202020200524
63488JS00003B/120